Cultural Studies of Science Education

Volume 19

The series is unique in focusing on the publication of scholarly works that employ social and cultural perspectives as foundations for research and other scholarly activities in the three fields implied in its title: science education, education, and social studies of science.

The aim of the series is to establish bridges to related fields, such as those concerned with the social studies of science, public understanding of science, science/technology and human values, or science and literacy. Cultural Studies of Science Education, the book series explicitly aims at establishing such bridges and at building new communities at the interface of currently distinct discourses. In this way, the current almost exclusive focus on science education on school learning would be expanded becoming instead a focus on science education as a cultural, cross-age, cross-class, and cross-disciplinary phenomenon.

The book series is conceived as a parallel to the journal Cultural Studies of Science Education, opening up avenues for publishing works that do not fit into the limited amount of space and topics that can be covered within the same text. Book proposals for this series may be submitted to the Publishing Editor: Claudia Acuna E-mail: Claudia.Acuna@springer.com.

More information about this series at http://www.springer.com/series/8286

Allison J. Gonsalves • Anna T. Danielsson
Editors

Physics Education and Gender

Identity as an Analytic Lens for Research

Springer

Editors
Allison J. Gonsalves
Department of Integrated Studies in Education
McGill University
Montréal, Canada

Anna T. Danielsson
Department of Education
Uppsala University
Uppsala, Sweden

ISSN 1879-7229 ISSN 1879-7237 (electronic)
Cultural Studies of Science Education
ISBN 978-3-030-41932-5 ISBN 978-3-030-41933-2 (eBook)
https://doi.org/10.1007/978-3-030-41933-2

This Springer imprint is published by the registered company Springer Nature Switzerland AG.
The registered company address is: Gewerbestrasse 11, 6330 Cham, Switzerland

Contents

Chapter 1
Introduction: Why Do We Need Identity in Physics Education Research?

Allison J. Gonsalves and Anna T. Danielsson

> An extreme culture of objectivity: a culture of no culture, which longs passionately for a world without loose ends, without temperament, gender, nationalism or other sources of disorder—for a world outside human space and time. (Traweek 1988, p. 162)

This is how anthropologist Sharon Traweek characterised physics culture, in her landmark book "Beamtimes and Lifetimes"—an account of the world of high energy physicists based on extensive field studies in laboratories in Japan and in the U.S. The "culture of no culture", Traweek argues, renders social categories of "physicist and physics community and physics culture" non-existent. Similarily, philosopher Sandra Harding has claimed that that the abstractness and the formality of physics need to be understood as "distinctive cultural features, not the absence of all culture" (Harding 1998, p. 61). However, getting sight of the cultural production of physics can be difficult and it is thus perhaps not surprising that when dealing with the issue of 'women in physics', that the gaze has more often been turned to the women than the physics. As such, over several decades, studies have documented differences between men and women in achievement or participation, or seeking social or psychological explanations for differences in physics engagement. This dualistic understanding of gender and its consequences for physics learning, engagement and educational research has long been challenged theoretically, but only recently have new perspectives on gender and physics been taken up in the field of physics education research. Feminist scholar Evelyn Fox Keller has suggested that binary understandings of gender and science render a situation where "any scientist who is not a man walks a path bounded on one side by inauthenticity and on the

A. J. Gonsalves (✉)
Department of Integrated Studies in Education, McGill University, Montreal, Canada
e-mail: allison.gonsalves@mcgill.ca

A. T. Danielsson
Department of Education, Uppsala University, Uppsala, Sweden
e-mail: anna.danielsson@edu.uu.se

A. J. Gonsalves, A. T. Danielsson (eds.), *Physics Education and Gender*,
Cultural Studies of Science Education 19,
https://doi.org/10.1007/978-3-030-41933-2_1

other by subversion" (Keller 1985, p.174). Responding to this, a recent epistemological shift in research on gender and physics education is turning our gaze away from documenting differences and rather moving towards understanding how gendered identities are constructed in physics learning and practice. For the past decade, we (Allison and Anna) have been researching students' experiences in physics education, and we have been exploring various uses of identity to understand these. We have seen shifts in how researchers approach explorations of gender in physics, and wish to document these new developments here, in this edited volume. The focus of this book will thus be on extending our theoretical understandings of identity to explore the construction of gender in the teaching and learning of physics, in and beyond the field of physics education research (PER).

1.1 Physics Education Research (PER)

PER is a research field that deals with the teaching and learning of physics, and is typically is considered a sub-field of physics rather than of education (Beichner 2009; Heron and Meltzer 2005). The field of PER is primarily concerned with university level teaching and learning of physics, even though some research groups reach across the whole spectrum of physics education from primary to university (e.g. University of Maryland). PER researchers tend to have a background in physics and it has been argued that it is appropriate for physicists to research university level physics education as they are the ones familiar with the complexities of university level physics coursework and the ones teaching such classes (Beichner 2009). At times, PER researchers have brought research methods from their investigations of the physical world with them into PER (Heron and Meltzer 2005). Research dealing with gender issues is a notable area of concern in PER—an interest motivated by the continued under-representation of women within the discipline. Recently, Scherr (2016) reviewed 400 articles from the past 10 years of publications in Physical Review Physics Education Research, and found that about 7% of those were concerned with issues of gender, and of those, 80% were focussed on performance gaps. Very few (3%) of the articles Scherr reviewed focussed on issues related to race, class, sexuality, disability or other social identities. However, a recent focused collection on "Gender in Physics" in *Physical Review of Physics Education Research* presented an emerging focus on identity work in physics learning. In particular were articles encouraging PER scholars to move away from binary models of gender that tend to focus on differences between men and women (e.g., Traxler et al. 2016), and rather to examine how identity might be a useful lens for understanding physics experiences (e.g., Gonsalves, Danielsson and Pettersson 2016). The use of identity frameworks in gender research is already well-established in science education research (e.g., Brickhouse 2001; Carlone and Johnson 2007), so we begin this book from the position that PER scholarship focused on gender issues may benefit from this promising orientation.

1.2 Repositioning Gender Research in PER

> The relationship between gender and science is a pressing issue not simply because women have been historically excluded from science, but because of the deep interpenetration between our cultural construction of gender, and our naming of science. The same cultural tradition that names rational, objective, and transcendent as male, and irrational, subjective, and immanent as female, also, and simultaneously, names the scientific mind as male, and material nature as female...Modern science is constituted around a set of exclusionary oppositions, in which that which is named feminine is excluded, and that which is excluded—be it feeling, subjectivity, or nature—is named female. Actual human beings are of course never fully bound by stereotypes, and some men and some women—and some scientists—will always go beyond them. But at the same time, stereotypes are never idle. (Fox Keller 1987, p. 279).

Since the 1980s, work of feminist theorists like Evelyn Fox Keller have compelled us to learn to count past two, that is, to challenge the dualisms that produce and reproduce men and women as different, and position them as naturally or unnaturally inclined towards masculinized subjects like physics. In the same period, empirical studies of physics cultures (such as Sharon Traweek's seminal anthropological work) also began to unveil how binary notions of gender are produced in physics. In 1990, Judith Butler gave us a language to begin to trouble the binary categories around which gender is constituted. The work presented in this book takes as a starting point that gender, like identity, is performative and fluid. Butler suggests that identity work involves a negotiation around subject positions (possible identities) that are simultaneously imposed and taken up. Applying various frameworks, all to some extent based in a fundamental notion of gender and identity as performative, the authors in this book approach gender not as a static trait that one possesses, but rather as something at is constructed between individuals in various social settings (like physics classrooms or labs). Thus, what it means to perform a feminine or masculine identity in these spaces can be context specific, and recognized differently in various situations. This approach to understanding gender in PER demands asking different questions about learning. For example, rather than asking "how do men and women learn these concepts differently?" we may ask "how are ideas about masculinity and femininity produced in these settings?" and then "how do people navigate these understandings in order to be recognized as competent in physics?"

This relational approach to gender demands that we also take into account identity and identity work. In past decades, identity has become a central theoretical concept in many disciplines, particularly science education. As this book will detail, identity frameworks have much to offer our understanding of gender in physics education research, yet identity has been undertheorized and underutilized in relation to physics learning. Frameworks that highlight identity work in physics can be used to explore how gender interacts with constructs like power, privilege, agency, discourse, positionality and inequity and how these are tied up in identity construction and trajectories into and out of physics. For instance, in her 2001 Journal of Research in Science Teaching article, Nancy Brickhouse argues that "in order to understand learning in science, we need to know much more than whether students have acquired particular scientific understandings. We need to know how students engage in science and how this is related to who they are and who they want to be." (p. 286) and advocates for the usefulness of a perspective on science education that consider learning as identity formation. From this perspective, identity is perceived as something we do rather than something we are (Carlone and Johnson 2007). Following Brickhouse's influential work, this perspective on science education has been developed by scholars such as Angela Calabrese Barton (e.g., Calabrese Barton 1998) Heidi Carlone and Angela Johnson (e.g., Carlone and Johnson 2007), and Louise Archer (e.g., Archer et al. 2012), and colleagues. Building on the theoretical work of Dorothy Holland, James Gee, Pierre Bourdieu, and Judith Butler, these science education researchers are studying gender by examining identity work. This important work has gained significant purchase in science education research communities, and the recent focussed collection of PRST-PER suggests that the PER community is taking notice. By applying these performative perspectives of gender and identity to physics learning, we can begin to see that not all identity or gender performances are equally feasible. In physics learning environments 'who' can be recognised as a certain 'what' can be limited by situational and structural constraints (including the body of the individual). As argued by Gonsalves et al. (2016) such a perspective 'helps us look more carefully at the complexities of gendered experiences in physics environments, rather than simply asking questions about what women need to succeed in physics' (p. 3).

This edited collection expands our understanding of gendered participation in physics from a *binary gender deficit model* (Traxler et al. 2016) to a more complex understanding of gender as performative and intersectional with other social locations (e.g., race, class, dis/ability, etc). The work presented in this book contributes to a growing scholarship using sociocultural frameworks to understand learning and participation in physics, and that seeks to challenge dominant understandings of who does physics and what counts as physics competence. Studying gender in physics education research from a perspective of identity construction allows us to understand participation in physics cultures in new ways. We are able to see how identities shape and are shaped by inclusion and exclusion in physics practices, discourses that dominate physics cultures, and actions that maintain or challenge structures of dominance and subordination in physics education.

1.3 New Perspectives on Gender and Identity in PER

The chapters offered in this book present new perspectives on understanding identity in PER while at the same time constructing a broad picture of the complexity inherent in *doing* physics and *doing* gender. Various perspectives on gender and identity will be explored via commentaries and empirical evidence emerging from a range of participants (in upper secondary settings, and higher education settings), employing a variety of research methodologies and analytic lenses. Several chapters take up examinations of the discursive practices that construct insider identities in physics. In Chap. 2 Louise Archer, Emily MacLeod and Julie Moote present data from interviews collected over 7 years from girls who had expressed aspirations to study physics. Archer and colleagues draw on concepts from Bourdieu to understand how girls move in, through, and out of physics. The data presented in this chapter, and its theoretical framing, help us to understand the gatekeeping practices present in the field of physics, and how these practices help to ensure the reproduction of the fields' elite status alongside the marginalization of women from the field. Louise Archer, Julie Moote and Emily MacLeod then present different data from the ASPIRES project in Chap. 4, this time with a focus on constructions of masculinity that permit a seamless or normalised trajectory into the study of astrophysics. Archer et al. once again draw on longitudinal data and a Bourdieusian framework to examine how interactions of capital, habitus and field work together to possibilise and normalise a male student's trajectory into becoming a physics student. This framework illuminates how notions of masculinity work with a discursive construction of cleverness that are normalised and demanded in physics, and how these work in positive ways for this student, but may operate to marginalise others.

New to the field of PER are perspectives on how intersectionality (e.g., Crenshaw 1989, 1991) can be taken up theoretically and operationalized in research on physics cultures. In Chap. 4 Angela Johnson provides an introduction to the framing of identity with an intersectional analysis in her chapter that explores intersectional physics identities, in higher education learning environments that seem to work well for women physics majors of colour. Unique to this work is her perspective that identity is not an individual experience, but rather a *feature of a social setting*. This opens up many possibilities for questions about how personal interactions, cultural features and structures in various settings can send messages about what kinds of people belong in those settings. The intersectionality lens Johnson applies here helps us to see how different kinds of people may experience the same setting differently depending on their various social identities and personal characteristics. Similarly, in Chap. 5, Diane Crenshaw Jammula and Felicia Moore Mensah present us with stories of students in physics labs that highlight the intersections of masculinity, femininity and racialized subjectivities, and how the alignment of White, middle class, masculine subjectivities with conventional physics afford male students the confidence to define what counts as physics in laboratory spaces. This chapter presents us with the innovative use of reflective journals as a data collection method, and an insider perspective as the lead author Crenshaw Jammula was also the physics class instructor.

Chapter 6 presents a different perspective on gate-keeping, this time considering the role that physics jokes have in constructing a discursive field that is accessible to some but inaccessible to others. In their chapter, Anders Johansson and Maria Berge explore the discursive construction of physics culture through lecture jokes in university quantum mechanics classes. These researchers draw on ethnographic data to explore questions about how physics lecture jokes may structure possibilities for students to identify with physics and as physicists, by constructing celebrated subject positions through jokes, which may do the discursive work of positioning students inside or outside of physics.

Two contributions to this collection move the field into very new theoretical territory with the introduction of sociomaterialist and critical disability frameworks. In Chap. 7 Marianne Løken and Margareta Serder take up Barad's (1999) notion of 'intra-action' with material objects to understand women's educational choices that lead them towards physics careers. This framing helps us to understand how materials, and students' intra-actions with them may play a role in gendered educational choices. This post-humanist perspective on gender and the material has been well-developed in the field of science and technology studies (e.g., Asberg and Lykke 2010), but is very new in the field of physics education research, and will be of interest especially to those involved in laboratory design and out-of-school-experiences intended to attract diverse youth to physics. Also providing new theoretical perspectives to consider is the work of Adrienne Traxler and Jennifer Blue presented in Chap. 8. Like Johnson, these PER researchers remind the reader that gender is never the 'whole story', but rather only one way to signal 'not-belonging'. Traxler and Blue draw on DisCrit--a recent synthesis of critical dis/ability and critical race theory--and crip theory, which studies the intersection of disability with LGBT identities to consider how we might begin to look beyond exclusively gender-focussed frameworks to understand identity work in physics.

Finally, in Chap. 9 Jaimie Miller-Friedmann presents the trajectories of successful women in physics through a narrative account of the strategies and tactics used by female academics who have had significant success in the field of physics. Miller-Friedmann's analysis demonstrates three significant experiences and identity negotiations that facilitated women's persistence in physics: 1) reliance on the self, 2) social support networks, and 3) the construction of a working class hero identity. This work presents suggestions for recruitment and retention of women in physics.

The book concludes with contributions for practitioners in physics education in higher education and upper secondary levels. Dimitri Dounas-Frazer discusses his positionality as a queer physics professor in higher education contexts and how this positionality informs his views on dualisms that shape social relations in physics. His commentary draws on lessons from the chapters in this volume that caused him to interrogate his professional and personal identities in relation to physics teaching, and pedagogical changes he has made in response to these reflections. Dounas-Frazer describes specific actions he has taken to "take gender seriously" in physics classrooms, including developing "accountability partnerships" with colleagues to support gender- and race-based equity in higher education physics contexts. Similarly, Christopher Gosling—a physics teacher in a secondary school in rural United States—interrogates his own positionality in relation to both physics and his physics students, in response to his reflections on the chapters in this book. Gosling provides his reflections on the chapters most salient to post-secondary physics environments, and highlights themes related to *gendered norms* in classroom practice, *cleverness* as a pre-requisite for success in physics, and gendered assumptions about *interest* in physics. He discusses how these themes have influenced his own teaching, and provides advice and solutions for practitioners in secondary physics education classrooms.

Studying gender in physics education research from a perspective of identity and identity construction allows us to understand participation in physics cultures in new ways. We can see how identities shape and are shaped by inclusion and exclusion in physics practices, discourses that dominate physics cultures, and actions that maintain or challenge structures of dominance and subordination in physics education. The chapters offered in this book will focus on understanding why researchers in PER can benefit from identity framings and its usefulness in various contexts with various learner or practitioner populations. This scholarship collectively presents us with a broad picture of the complexity inherent in *doing* physics and *doing* gender, in the "culture of no culture".

References

Archer, L., DeWitt, J., Osborne, J., Dillon, J., Willis, B., & Wong, B. (2012). Science aspirations, capital, and family habitus: How families shape children's engagement and identification with science. *American Educational Research Journal, 49*(5), 881–908.

Åsberg, C., & Lykke, N. (2010). Feminist technoscience studies. *European Journal of Women's Studies, 17*(4), 299–305.

Barad, K. (1999). *Agential realism: Feminist interventions in understanding scientific practices* (pp. 1–11). The science studies reader.

Beichner, R. J. (2009). An introduction to physics education research. *Getting Started in PER, 2*(1), 1–25.

Brickhouse, N. W. (2001). Embodying science: A feminist perspective on learning. *Journal of Research in Science Teaching, 38*(3), 282–295.

Butler, J. (1990). *Gender trouble: Feminism and the subversion of identity*. New York: Routledge.

Calabrese Barton, A. (1998). Teaching science with homeless children: Pedagogy, representation, and identity. *Journal of Research in Science Teaching: The Official Journal of the National Association for Research in Science Teaching, 35*(4), 379–394.

Carlone, H. B., & Johnson, A. (2007). Understanding the science experiences of successful women of color: Science identity as an analytic lens. *Journal of Research in Science Teaching, 44*(8), 1187–1218.

Crenshaw, K. (1989). Demarginalizing the intersection of race and sex: A Black feminist critique of antidiscrimination doctrine, feminist theory and antiracist politics. In *University of Chicago Legal Forum* (Vol. 140, p. 139).

Crenshaw, K. (1991). Mapping the margins: Identity politics, intersectionality, and violence against women. *Stanford Law Review, 43*(6), 1241–1299.

Gonsalves, A. J., Danielsson, A., & Pettersson, H. (2016). Masculinities and experimental practices in physics: The view from three case studies. *Physical Review Physics Education Research, 12*(2), 020120.

Harding, S. G. (1998). *Is science multicultural?: Postcolonialisms, feminisms, and epistemologies*. Indiana University Press.

Heron, P. R., & Meltzer, D. E. (2005). The future of physics education research: Intellectual challenges and practical concerns. *American Journal of Physics, 73*, 390–394.

Keller, E. F. (1985). *Reflections on science and gender*. New Haven/London: Yale University Press.

Keller, E. F. (1987). On the need to count past two in our thinking about gender and science. *New Ideas in Psychology, 5*(2), 275–287.

Scherr, R. (2016). Never mind the gap: Gender-related research in physical review physics education research, 2005–2016. *Physical Review Physics Education Research, 12*(2), 020003.
Traweek, S. (1988). *Beamtimes and lifetimes*. Cambridge: Harvard University Press.
Traxler, A. L., Cid, X. C., Blue, J., & Barthelemy, R. (2016). Enriching gender in physics education research: A binary past and a complex future. *Physical Review Physics Education Research, 12*(2), 020114.

Chapter 2
Going, Going, Gone: A Feminist Bourdieusian Analysis of Young Women's Trajectories in, Through and Out of Physics, Age 10–19

Louise Archer, Emily MacLeod, and Julie Moote

2.1 The Exclusion of Women and Femininity from Physics

The 'gender problem' in physics is a long-standing and widely recognised issue. Women remain under-represented in post-compulsory physics (e.g. Smith 2010a, b, 2011), despite decades of interventions aimed at improving the gender profile of physics (Darke et al. 2002). These differences are not the result of inequalities in female attainment in the subject (Haworth et al. 2008; Smith 2011; Tan et al. 2013; Tytler et al. 2008). Rather, attention has been drawn to the masculine *culture of science* (Harding 1998), and the multiple ways in which this disadvantages and excludes women (Blickenstaff 2005), through explicitly gendered curricula and representations of the subject (e.g. Baker and Leary 1995;) which girls struggle to find relate to (Calabrese et al. 2006; Calabrese Barton et al. 2008; Haussler and Hoffmann 2002), to the gendered biases of teachers (Carlone 2004), and these unconscious understandings people have developed in which physics is seen as being 'for boys', such that girls receive less encouragement from others to pursue the subject (Mujtaba and Reiss 2013). Indeed, even those women who pursue the subject at degree level find it hard to reconcile their femininity with a legitimate physics identity (Danielsson 2012; Gonsalves 2014).

L. Archer (✉) · E. MacLeod · J. Moote
UCL Institute of Education, London, UK
e-mail: l.archer@ucl.ac.uk; e.macleod@ucl.ac.uk; j.moote@ucl.ac.uk

A. J. Gonsalves, A. T. Danielsson (eds.), *Physics Education and Gender*, Cultural Studies of Science Education 19,
https://doi.org/10.1007/978-3-030-41933-2_2

2.2 A Feminist Bourdieusian Conceptual Lens

Despite Bourdieu's relative lack of interest in gender in this work (cf. Bourdieu 2001), feminists have still found his theoretical tools useful and productive, particularly for reconceptualising identity – as simultaneously part of (produced by and incorporating) the social world (Bourdieu 1984, 1990; Bourdieu and Passeron 1972; Bourdieu and Wacquant 1992). As Adkins explains, 'the notion of the subject as not simply engaged with the world, but *in* the world, is one which has great appeal to feminists' (Adkins 2004: 10, emphasis in original). Bourdieu's theory breaks down the Cartesian dualism and provides a framework in which gender can be understood as not just a product of the mind/ consciousness. That is, his work provides a conceptual framework for understanding gender whereby 'mind and body, thought and action, are indissolvable' (Adkins 2004, p. 11).

Drawing on Moi (1991, 1999), Adkins (2004) argues that gender is best conceptualised as part of the general social field, being 'extraordinarily relational, with a chameleon-like flexibility, shifting in importance, value and effects from context to context or from field to field'. In other words, gender is both 'dispersed across the social field and deeply structuring of the social field' (Adkins 2004, p. 6). For feminists, Bourdieu's emphasis on embodiment is also highly congruent with feminist approaches to identity, as epitomised by his conceptualisation of the habitus (the internal framework of socialised dispositions which is both structured by and structuring of experience) as being both an embodiment of the social world (hexis) and a socialised body that shapes the social world.

While feminists have generally found little use for *Masculine Domination* (2001) per se, Bourdieu's sole text focusing on gender, there is a treasure trove of feminist extensions and adaptions of his main ideas. For instance, McNay (1999) extends Bourdieu to understand gender as 'a lived social relation which will always involve conflict, negotiation and tension' (Adkins 2004: 11), in which experience is always relational but not foundational. Lawler (2004) has also provided a feminist Bourdieusian analysis of how gendered and classed identities are conferred on subjects via cultural authorization by the media.

2.3 The Aspires/ Aspires2 Study

For our analysis, we draw on data that were collected as part of the Aspires/ Aspires2 project – a 10 year, mixed methods study of children's science and career aspirations from age 10–19 that was funded by the UK Economic and Social Research Council. Our wider dataset comprises large-scale national surveys of a cohort of students as they progress through primary and secondary schooling, combined with in-depth longitudinal interviews with a subset of young people and their parents (e.g. see Archer et al. 2017). However, for this chapter, we focus on a subset of interviews that were conducted with seven young women: Danielle, Davina, Hannah, Kate,

Table 2.1 Demographic details of the seven young women

Name	Science GCSEs	Science A levels	Post-18 destination	Social class	Ethnicity	Gender identity at age 18
Danielle	Double science	–	Sociology degree	Working-class	White British	Female
Davina	Triple Science	Maths, physics, chemistry	Gap year (with offer for Chemistry degree next year)	Upper middle-class	White British/ European	Female
Hannah	Triple Science	Chemisty, maths, further maths, physics	Physics degree	Upper middle-class	White British/ North American	Female
Kate	Triple Science	Biology, chemistry, physics, maths	Natural sciences degree	Upper-middle-class	White British	Female
Mienie	Triple science	Chemistry, Maths, Physics	Gap year (before taking up offer to do Chemistry degree)	Middle-class	South Asian	Female
Thalia	Triple science	- (*started but did not finish physics A level*)	Japanese studies degree	Middle-class	White British	Other
Victoria	Triple science	Maths, DT (*started but did not finish physics A level*)	HE foundation engineering course (with view to do electrical engineering degree following year)	Middle-class	White British	Female

Mienie, Thalia and Victoria (see Table 2.1). These young women were selected from the wider dataset on the basis that by 16 (Year 11,. in the English school system), they had all expressed an aspiration to study Advanced level (A level[1]) Physics. Interviews were conducted at five time-points; Year 6 (10/11), Year 8 (11/12), Year 9(12/13), Year 11 (14/15), and Year 13 (age 17/18).

As can be seen from Table 2.1, six of the girls went on to study advanced level physics (Davina, Hannah, Kate, Mienie, Thalia and Victoria) but the seventh, Danielle, was dissuaded by her school from taking the subject. Four of the young women successfully completed physics A level (Davina, Hannah, Kate and Mienie) but two were deselected from the course by their school part way through their studies (Thalia and Victoria). Only one young woman (Hannah) went on to study for a physics degree, although five of the seven young women pursued other science/ STEM-related routes post-18. Using our analytic lens, we now explore the young

[1]Advanced Level / A level examinations are the 'gold standard' post-compulsory academic qualifications that are studied over 2 years with final examinations taken at age 18. A levels are the most usual qualifications that provide entry to university degree courses.

women's accounts in more detail to try to understand the processes at work which facilitate, or prevent, their physics trajectories.

2.4 Impossible Female Physicists – Girls Denied Entry and/or Debarred from A Level Physics

There were three girls (Danielle, Thalia and Victoria) who at age 15/16 had wanted to study A level physics but who ended up not achieving this goal. All three girls were interested in the subject:

> I love Physics [...] I like it because Physics is kind of like Maths and English and it's one ... like it's ... if you were sitting in ... let's say [...] if you were sitting a Physics exam you can read the question and work ... do the equations and work it out, even if you knew nothing about what it was all about [...] That's what I like about Physics. It's kind of like common sense. (Danielle, Y11)
>
> I was doing Psychology, but then I decided ... I was talking to my Physics teacher about Physics, because I really like Physics and he said he did think I'd do fine, because I didn't think I'd do well in Physics, but he kind of was like no, you'll do all right. You'll be fine, so then I changed Psychology to Physics. (Thalia, Y11)
>
> I do enjoy Physics [...] I just like knowing how things work, what's going on out there in space and everything. Just I like knowing what's – why things do what they do. (Victoria, Y11)

In other words, none of these girls seemed to suffer from a lack of interest, aspiration or inspiration – as is so often assumed by initiatives aimed at encouraging girls into physics. Rather, in each case, as we now discuss, we interpreted their inability to continue as due to stringent gatekeeping – and practices of debarring – enacted by the education system.

Prior to taking her GCSE examinations (the national examinations taken at age 16 in England), Danielle had told her school that she wanted to study A level physics. In her Y11 interview she told us that the school had explained that acceptance onto the A level course would be dependent on achieving a B grade in science in her upcoming GCSE examinations:

> I was going to take Physics but I've got to wait and get my results from my exams back because if I don't get a B, I can't do physics (Y11)

She chose psychology as a back-up subject, although maintained a preference for physics ("if I can get it"). In the end, after not attaining a B grade in Physics GCSE, Danielle studied Sociology A level (along with English, Media and Health & Social Care) because she did not attain the B grade in maths that was required for entry to Psychology A level. After her results, Danielle had asked her school again to be considered for A level Physics and although the teacher initially consented to possibly give her a 'trial', Danielle was informally persuaded that she would probably find the course too difficult and, in end, did not take up the trial offer.

> I asked my teacher if I could do physics, even though I had a C and she said 'yeah' but then I got put off because apparently it's really hard (Y13)

Danielle reflected that "at the time I was upset" but consoled herself that it was probably the 'right' decision "cos I know so many people that are like failing science [...] like *really* failing" (Y13 interview).

Unlike Danielle, while Thalia and Victoria were allowed entry to physics A level, both were expelled from the course by their respective schools at the end of their first year of study and were not allowed to progress with the subject. In both of their cases, the reason for this debarring was on the basis of insufficiently high attainment to date on the course. As Thalia put it, "I didn't do very well in physics" and got a low mark on her end of year mock examination. Victoria similarly 'failed' the first year mock examination and explained that her teachers had advised her that she drop out altogether and not return to the college at the start of the next academic year:

> I was on the cusp of being told like 'sorry, no you can't come back' and they [teachers] strongly advised not coming back. But I came back and now I'm [predicted as] going to get an A in DT [Design Technology], I'm going to get a B in Politics and – not so great but – I'm going to get a D in Maths

Based on the grades she obtained at the end of Year 13, Victoria entered a one year engineering foundation course, from which she planned to go on to an electrical engineering degree. She explained that she made her choice of engineering course and degree pragmatically, via a 'process of elimination', according to her grades and the course entry requirements). Victoria thus agentically negotiated the schools' 'advice' that she not continue, to carve out a route towards an engineering degree, in order to achieve her long-held aspiration. Importantly, Victoria possessed a range of cultural, social and science-related capital to draw on that helped her navigate this trajectory – and which supported her to mitigate the pedagogic work undertaken by the school, which encouraged her to 'leave'. However, we also note the imprint of the field on her habitus in that she blamed herself for her 'failure' to continue and did not question the legitimacy of the decision. For instance, she did not question the attainment bar as an arbiter of who is allowed to continue with the subject, nor did she question the arbitrary 'jump' in 'difficulty' from GCSE to A level. Rather, she located the issue as her inability to 'click' with the subject:

> I didn't click with any of the Physics. I feel like that was always a bit hopeless, but I got a B at GCSE, but that compared to A level it's just not even comparable

Applying our conceptual lens, we interpret Victoria's quotes as exemplifying the effect of long-term pedagogic action, which reproduces physics as an elite subject, whereby the practices that are key to maintaining this eliteness (e.g. performances of high attainment as both a pre-requisite for acceptance on to the A level and for retention while studying the course; grade severity and arbitrarily 'difficult' content at A level; the requirement for physics on higher status engineering degrees) are not questioned but are accepted as doxa. As a result, 'failure' is ascribed to the individual student. Indeed, both Thalia and Victoria accepted the legitimacy of the decision to debar them from the course. From a Bourdieusian analysis, this might be read as an example of practical faith – whereby practices of debarring are designed to obtain 'native compliance' (Bourdieu 1990, p. 68). We suggest that the example practices noted in our data set are not uncommon, as national government data indicates that physics A level typically demands higher level entry qualifications (typically A/A∗) compared to other subjects (OfQual 2017). Moreover, several analyses suggest that physics A level is marked more harshly than other subjects, making it more difficult to achieve a high grade and, in turn, encouraging schools to enact more stringent gate-keeping practices in the subject (Thomson 2015; Tracy 2016a, b). Such practices are key to the reproduction of the doxa that physics is 'difficult' and only for the 'clever'. These practices operate as a form of pedagogic work, not only controlling who is allowed to enter and remain on A level physics courses, but also – as we discuss next – cultivating the habitus, such that many students come to self-exclude from continuing with physics.

2.5 Self-Exclusion and the Cultivated Habitus – Young Women Who Leave Physics After A Level

Three of the seven young women – Davina, Kate and Mienie – completed A level physics but did not choose to continue the subject post-18. As we have written previously (Archer et al. 2017), compared to girls in the wider data set, these young women were all highly 'exceptional'. Not only were they unusual given the very low proportion of girls who study physics, but they were also distinctive on account of possessing high levels of (cultural, social and specifically science capital), achieving very highly in academic terms, and on account of their relative comfort with 'being different' from the majority of girls with regard to their gender identity. In particular, they did not conform to 'girly' popular femininity. In short, they were, as Davina put it in her Year 11 interview, 'not like your average person':

> Maybe not a lot of people are good at Science and maybe that's kind of the *beauty* of being someone that is good at Science, I guess. The fact that maybe you're not like your average person (Davina, quoted in Archer et al. 2017, p. 100).

However, over the course of studying A level physics, these three young women came to the view that degree level physics was not 'for me'. Unlike Danielle, Thalia and Victoria, they were not debarred from continuing, indeed, they achieved grades (e.g. Davina, A, Kate, A∗), which would have enabled them to apply to study the subject at degree level. Rather, we now discuss, they self-excluded, which we interpret as due to pedagogic work conducted by the subfield of school physics, enacted through the notion of the 'effortlessly clever physicist'.

As we have noted among younger female students (Archer et al. 2013) and indeed among secondary students more generally (Archer et al. 2017), physics has a long-standing association with notions of 'cleverness' and is widely identified as being the 'hardest' of the sciences. The girls who continued with A level physics were no exception and notions of cleverness (as 'measured' through the metric of attainment) were frequently brought up within the girls' reflections regarding whether they felt they could continue further with physics, or not, post-18. For instance, Kate described how despite having 'always liked' physics, she had always questioned whether it might be 'for me' due to the need to continually produce high levels of attainment to remain viable as a physics student:

> I think I've like always liked physics, but always thought it was quite hard, so maybe not for me, but then I was like 'oh well I'm doing okay at school. I might as well keep it up as long as I can' (Kate, Y13)

Kate's insecurity was particularly striking as she recorded the highest GCSE and A level attainment out of all the students we followed in the study, achieving 10 A∗s at GCSE and 3 A∗s and an A grade at A level. We interpret her concern – that physics is "quite hard, so maybe not for me" – as exemplifying the effect that the subject (and its concerted pedagogic work) has had on her habitus, notably inculcation of the expectation that only the 'cleverest' and highest attaining students are legitimate participants in the subject. Hence, despite her clear aptitude, interest and credentials, Kate excluded herself from the possibility of studying a physics degree, worrying that it would be 'too hard':

> I wouldn't do like a straight physics degree, because it would be too hard. Like I think I'm just a bit put off by thinking that it would be really hard. […] So yeah, I think what put me off doing straight physics was that I think it's too hard and what put me off straight Biology is I'd quite like to do some physics as well. (Kate, Y13).

Davina's identification with physics – and the extent to which she considered herself a viable physicist – shifted over the years due in no small part, we would argue, to dominant associations of the subject with a particular notion of cleverness. Davina was a student with similarly high attainment (for instance, in her GCSE examinations she gained 8 A∗s, which included maths and sciences, 2 As and a B grade) and was studying maths, chemistry and physics at A level (all of which she achieved at grade A). When we interviewed her between the ages of 12 and 16, she had identified strongly with physics. As she reflected at age 18, "like definitely I was more like a physics… person". However, over the course of studying A level physics, despite her continued interest in the subject, Davina became less sure that degree-level physics was 'for her' and eventually applied to study for a chemistry degree instead. She described this process of negotiation at some length in her interview, which as the following extract shows, centred around the question of 'cleverness':

> I mean certainly if someone said 'do you think you're clever enough to do physics at university?', I would say definitely not, most definitely not … like no way I could do physics at university […] I mean I guess I'm probably smart enough to like get the A level, and then I don't think that necessarily means that I'm actually like that *good* at physics, if you know what I mean? (Davina, Y13, emphasis added)

As we discuss in Archer et al., (2020), most of the young women (whether they continued with the subject or not) struggled to recognise themselves as being 'good at physics', irrespective of their actual attainment (Mujtaba and Reiss 2013). Moreover, we interpreted the young women's accounts as highlighting a specific configuration of physics cleverness, one that must be 'natural' or 'effortlessly' produced in order to be legitimate. That is, one is either "clever enough", or not, to take the subject further – rather than, for instance, being able to continue on the basis of hard work and application.

Indeed, Davina described how she usually understood concepts and content more quickly and easily in the other sciences ("for most things in science I do tend to understand them like first time"), but she felt that this was not the case with A level physics ("in physics, I don't"). Mienie expressed a similar sentiment:

> I mean physics is really hard but I enjoy physics even though it is quite hard. I have to work quite hard for it …yeah, I don't know why but it's, I don't know, because some things come naturally to you and physics, it did, but then…

That is, the young women were concerned by their need to 'work quite hard' to understand A level physics and interpreted this as evidence that they were not authentic or viable degree level physicists. This notion was further cultivated by their perception that some peers (whether 'real' classmates, in Davina's case, or 'imagined' by Kate) did not have to exert a similar effort:

> Part of the reason why I'm maybe putting myself down slightly is probably because I'm comparing myself to people who are just kind of like … you know kind of again pretty much breezing through and getting like you know 80% or whatever. And then I'm there like trying really hard and getting less than that. (Davina, Y13).

Indeed, all three girls felt that they had to work much harder in physics than the other sciences – but rather than interpreting this as (for instance) the potential result of a more challenging curriculum, harder marking or more variable teaching, they located the 'fault' within their own essentialised capabilities (that is, being 'not clever enough' to study physics at university). Although Mienie did hint at the issue of teacher quality at other points in her interview, as the following point exemplifies, this was underscored by her description of how the teacher enacted pedagogic work by reproducing the notion of physics as not just 'hard' but is even potentially 'unknowable' by the students:

> Um, so for Physics I have two different teachers and the one particular teacher she, her teaching style is very not good. [Int: Oh] I'm just going to say not good. …Um, so yeah she, I mean she admits it herself sometimes that, sometimes she'll be explaining something and she'll just say 'oh if I was you, I wouldn't understand it myself' and I would think, yeah. (laughs) (Mienie, Y13)

We interpret these young women's accounts as signalling the cultivated physics habitus, in which a legitimate physics identity is aligned with 'effortless achievement' (Bourdieu and Passeron 1977). The notion of 'effortless' achievement has been identified by feminists as being a gendered construction that is aligned with masculinity, whereas attainment via 'effort' is associated with femininity (e.g. see Francis and Skelton 2005). Indeed, Carlone's (2004) study shows how powerfully these associations are within the context of advanced physics courses, such that teachers attributed boys' attainment to 'raw talent', whereas they explained away girls' (higher) attainment as achieved via 'plodding diligence' (and hence as not being produced from 'natural' aptitude for the subject).

We read the young women's accounts as showing traces of the symbolic violence that is inculcated within the habitus through their physics socialisation. As Bourdieu and Wacquant explain, symbolic violence is "the *violence* which is exercised upon a social agent with his or her complicity" (1992, p. 167, italics in original). That is, the girls attributed 'blame' and failure to themselves and self-excluded themselves from the possibility of taking the subject further. Indeed, despite attaining a top grade in A level physics, Kate said self-deprecatingly in the interview, "I just don't really understand it that well". When probed as to which of her classmates understood the content better, she could not identify anyone, but conceded that her reference point (for the legitimate 'naturally' clever physics student) was "maybe imagined" (Kate, Y13). We interpret this notion (of the imagined, effortlessly clever physicist) as hinting at the pedagogic action underlying such practices – namely, the myriad of everyday acts and practices that support the reproduction of the elite status of physics by restricting the entry and retention of all but a privileged few. That is, we suggest that the students' accounts hint at pedagogic work which is undertaken within the teaching and learning of A level physics which inculcates students to accept that physics is not just 'hard' but is 'too hard for all but the 'natural', effortlessly clever (male) physicist – and that the propagation of this fantasy is an integral part of the reproduction of the elite status of the subject.

In other words, we suggest that the notion of the 'effortlessly clever physicist', which is cultivated through a range of practices enacted within school physics and the wider media (as epitomised by the character of Sheldon in popular US TV comedy, *The Big Bang Theory*), seemed to play a key role in deterring even highly able, qualified and interested young women from seeing post-18 physics as appropriate and attainable. Rather than being debarred, we understand the doxa of the 'effortlessly clever physicist' as working silently and perniciously to make many students question their own legitimacy as someone who is 'actually good' at the subject and hence viable as a degree candidate, irrespective of their interest, enjoyment and attainment in the subject. Thus symbolic domination is achieved through the self-regulation of the cultured habitus. As Jenkins writes, pedagogic work aims to produce within the habitus 'dispositions which generate 'correct' responses to the symbolic stimuli emanating from agencies endowed with pedagogic authority" (2006, p. 107) – such as the notion of who is 'clever enough' to continue with physics. This process is both effective and powerful:

> The legitimate culture becomes experiences as an axiom, a *fait accompli*: Children all too soon stop asking 'Why?' Exclusion works most powerfully as self-exclusion (Jenkins 2006, p. 107).

Writing in the context of the production of working-class 'taste', Bourdieu argues that the working-class often make a virtue out of a necessity ('that is to refuse what is categorically denied and to will the inevitable' – 1990), 'inducing 'choices' which correspond to the condition of which it is a product' (Bourdieu 1984, p. 175). We suggest that this account seems to have some purchase for explaining the self-exclusion of Davina, Kate and Mienie from post-18 physics, but how does it account for Hannah, who – as discussed next – went on the study for a degree in physics?

2.6 Last Woman (Physicist) Standing – Hannah

Hannah was the only girl who, after completing A level physics, applied for a physics degree. She recorded high levels of attainment 7A∗s (inc. maths & science), 3As and a B at GCSE, and 4 A∗s at A level (in chemistry, physics, mathematics and further mathematics), although these were not notably higher than Kate, Davina and Mienie (who took physics A level but who did not continue further with the subject). However, Hannah did stand out from her peers in that she did not describe physics as particularly 'hard':

> Well parts of it [physics] are difficult but if you compare that to English I'd say physics was easy. [Interviewer: Right, yeah] Not easy but like easier. It's quite nice, cos if you just understand like the basics, you can pretty much figure out everything from there (Y13)

As we discuss in Archer et al., (2020), Hannah was also the only girl to assert the view that she might be 'good at physics', although notably she still did not align herself with the notion of the effortlessly clever physicist, claiming that she does not "breeze through" but rather has to "work to understand things":

> Well I'd like to think at least that I am good at physics. But not like breeze through it, you have to still like work to understand things. So probably like in the middle of that. There'll be people who like completely breeze through it – I'm not one of them (Y13)

Her views thus echoed those of Davina, Kate and Mienie – who described having to 'work' at the subject – although unlike Hannah, the others interpreted this requirement of effort as signalling that they should not continue with the subject. We interpret Hannah as negotiating a tricky identity tightrope – in which she produced a self-identification as being 'good at physics' (which is dominantly aligned with masculinity) but positioned this as being achieved via feminised diligence (having to "work to understanding things"). We suggest that this combination enabled her to maintain an intelligible femininity and identify as a viable, although not necessarily a dominantly authentic/ legitimate, physicist.

So how and why does Hannah end up pursuing a physics degree when so many other (equally well qualified and interested) young women do not? On one level, a Bourdieusian lens provides a helpful steer by alerting us to the importance of capital for enabling Hannah's physics trajectory. While the other young women who took the A level also enjoyed substantial economic, cultural, social and science-related capital, Hannah undoubtedly possessed the most specifically *physics*-related capital. For instance, Hannah engaged in a high volume and wide range of science (and physics)-related informal science learning activities over the years, including regularly going on the 'IFLS' website and reading about new developments in physics via a range of physics media:

> Physics came first cos I think … I think I read a book … I can't remember, it might have been Higgs [about the Higgs-Boson particle], and like I got a subscription to New Scientist for my birthday and I just started reading them. And I was like 'Oh that's actually pretty cool' (Hannah, Y13)

In particular though, Hannah had the most substantial physics-related social capital out of any of the students. She had several family members who were physicists and ended up going to the same university to study physics as her older brother. Such family pathways can foster not only an awareness and desire for particular options but can also provide pragmatic support and mitigation of risk, in that these trajectories constitute 'safe', known routes. For instance, in her Year 11 interview Hannah described the 'reassurance' that she had got from having an older brother who had done a postgraduate physics degree, and whose girl-friend (at the time, the couple later married) was a nuclear physicist and constituted an important source of information and support. In her Year 13 interview, Hannah also described the rich social capital that she derived from knowing the couples' many friends (whom Hannah had socialised with) who were also physicists:

> So I talked to her [brother's girlfriend, who is a physicist] quite a lot, cos she also wanted to do Medicine. [She talked] about like what you need to do to either ... well for Physics and for Medicine what you need to do like. And like what the jobs consist of [...] I was quite impressed with my brother's girlfriend cos she did quite like ... well I don't know she just seemed quite cool, cos her job was cool [...] I wouldn't mind doing that. (Y11).
>
> Yeah, he [brother] really likes it. And then I started talking to him about it. ... He did a PhD thing at Manchester, and he said that was amazing, which is one of the reasons I looked at it ... cos I guess it's a bit reassuring if you know somebody's done it and they still liked it afterwards (Y13).

Outside of her family, Hannah also felt that it had been important for her to have a best friend who shared her interests:

> She [friend] has the New Scientist and we discuss that as well. [...] So it's definitely helped ... because if you don't have someone sharing your interests it's really hard to like talk about them, which is kind of hard. (Y13).

We suggest that Hannah's impressive physics capital is one of the key factors in understanding her exceptional trajectory into post-18 physics. However, it is not clear why capital alone would necessarily result in Hannah transgressing gendered norms. It is here that the explanatory power of our Bourdieusian conceptual framework struggles and has to work harder. Most of Bourdieu's work was concerned with explaining social reproduction (predominantly in relation to social class) – it is less obvious to what extent it can explain female physics students, as an example of those who 'go against the grain' of social reproduction. The closest example we can find in Bourdieu's work is the example of working class 'survivors' (those who, like Bourdieu himself, attain social mobility through the education system), discussed by Bourdieu and Passeron (1977). This notion is usefully and productively extended by Reay, Crozier & Clayton (2009) in their study of working-class students who attended an elite university. Using a Bourdieusian lens, Reay et al. empirically unpicked how, within the unfamiliar setting of elite Higher Education, the working class habitus can produce a range of 'creative adaptions and multifaceted responses' (Reay et al. 2009: 1103). These adaptions were enabled through a constant 'fashioning and refashioning of the self' through reflexive interactions with the field (Reay et al. 2009, p. 1111).

Bourdieu suggested that a habitus which encounters an unfamiliar field (such as the working class student in an elite Higher Education setting) can result in 'a habitus divided against itself' (Bourdieu 1999b). However, arguably for Hannah, the field of physics was not as 'unfamiliar' as elite Higher Education might be for working-class students. Indeed, Hannah had been studying the subject for 7 years by the time of her last interview. Moreover, as we argued above, the field had been cultivating Hannah's habitus over time, such as to produce a notable degree of alignment and practical faith. Rather, we propose that we might better understand the A level physics girls as displaying examples of a dissonant habitus that results from intersectional encounters, in which Hannah, as an upper middle-class white young person, is both a member of the dominant ethnic and social class (and hence 'entitled' to continue with an elite subject, such as physics) while also occupying a position of subordination by dint of her gender. That is, for Hannah there is no 'sudden shock' of immersion into an alien field (as was experienced by the students in Reay et al.'s study). Indeed, some aspects of the field (e.g. the elite class alignment of physics) may be culturally familiar for Hannah. Moreover, we suggest that Hannah's prolonged experience of, and immersion in, physics (both in and out of school) had cultivated a degree of alignment between her habitus and the field. However, Hannah's habitus did have to work hard to resolve tension and dissonance between her *femininity* and the field of physics (as a field that is dominantly coded as masculine).

In earlier work we noted that, among younger (e.g. age 10–14 year old) science-keen girls, 'bluestocking' performances of femininity were much more common than 'girly' performances of femininity (Archer et al. 2012a, b). This pattern exacerbated over time, such that by A level, none of the six girls who studied physics self-identified as 'girly'. Rather, the young women's accounts conveyed a gradual alignment over the years, in which performances of femininity were increasingly regulated and downplayed, as exemplified by Hannah and Davina:

> I'm not … not particularly feminine … well I do ballet and everything, so that's quite feminine I guess. But not particularly like …[…] Um … well […] I'm just more comfortable in jeans […] I've cut my hair really short … […] Like really really short (Hannah, Y11).

> I wouldn't say I'm a particularly feminine person at all. I mean you know like I swear quite a lot (laughs) […] I swear like a sailor, it's ridiculous. You know I don't … first of all I don't really dress particularly feminine, like I tend to wear jeans and like band t-shirts and hoodies and stuff, and I wear boys' like skater shoes. So I mean yeah I'm not … I don't have a particularly feminine voice either … and I think well so what? – like there's nothing wrong with that, it's just like that's just what I am. (Davina, Y11).

Hannah and Davina talked about feeling different from many of their female peers, a sentiment that is mirrored by the working-class students in Reay et al.'s study, who also recounted long-standing feelings of difference from their working-class families and peers (in this case, on account of their academic dispositions). For instance, while in her earlier interviews, Hannah described feeling different to other girls, over time she developed this into a point of value – describing herself by the age of 16 as being "proud to be different" to other girls (see Archer et al. 2017).

While there may be some echo here with Bourdieu's discussion of how the working-class often make a 'virtue out of a necessity' (1984, p. 175), we suggest that this case differs due to the differential power relations in play. That is, while it became necessary for Hannah to eschew popular femininity in order to be intelligible as a physicist (see Archer et al. 2017) – because 'girly' femininity is dominantly configured as the antithesis of 'serious', masculine, rational physics (Francis et al. 2017) – unlike the working-classes discussed by Bourdieu, who were finding ways to 'make do' with their subordinated social position, Hannah pursued an elite trajectory. In this respect, while Hannah may have needed to negotiate the 'loss' (or suppression) of some valued aspects of femininity, as a high status route aligned with masculinity, physics also offered her a chance to 'get on', rather than 'make do'.

We thus interpret Hannah's gender performances as strategic and required adaptions to the field, aimed at resolving her gender dissonance to enable a successful trajectory, but also cultivated as part of her unconscious socialisation by/into the field. Indeed, Bourdieu recognised that habitus can be changed by field, being 'restructured, transformed in its make-up by the pressure of the objective structures' (Bourdieu 2005: 47), such that habitus constitutes an 'open system of dispositions' that are 'endlessly transformable' (Bourdieu 1990, p. 116). In this respect, we suggest that the young women who continued with physics may exemplify the process described by Puwar (2004, p. 128), in which they came to 'partially mirror and clone the self-image of the hegemonic norm'. Moreover, we suggest that Hannah's physics aspirations were arguably in line with Bourdieu's logic of capital accumulation (which he considered to be a key driver of the habitus), in that she was investing in a trajectory that offered status and capital.

Hannah's lived experience of femininity may also constitute a resource and form of capital that actively supported her physics trajectory. For instance, Reay et al. (2009) discuss how working-class students' experiences of living subordination can lead to resilience, self-reliance and determination, which support and facilitate the students to be successful in their elite Higher Education trajectories. Likewise, it is possible that Hannah's gendered experiences (both generally and specifically in relation to being a woman in the male-dominated field of physics) have provided her with 'grit', resilience and self-sufficiency that have supported her onward progression. Reay et al. suggest that the development of such qualities may be key for enabling the successful trajectories and experiences of 'non-traditional' students within elite fields:

> ... these students are Bourdieu and Passeron's (1977) working-class exceptions that prove the rule. Their combination of highly developed academic dispositions and reflexive habituses generate opportunities and academic success. (Reay et al. 2009, p. 1115).

These embodied resources may be developed through reflexivity – which Bourdieu suggests is key for the transformative habitus (or as Adkins 2004, p. 10 explains it, 'Bourdieu will always break with his main theoretical principles and will see the possibilities for social change when a conscious or thinking mastery of the principles of the habitus can be gained'). Arguably, Hannah – and indeed the other girls who completed Physics A level – displayed reflexive habituses. For instance, Hannah recounted how she was highly aware of being in a gender minority (as a girl studying physics and further maths in a co-educational school) and had thought (and internally debated) the issue extensively when choosing her courses.

> … because I knew I was going to be the only girl, I was getting really worried because then I was like … if I'm the worst in the class it's just going to be like extra pressure because you don't want to … I guess being a girl can put extra pressure on you, cos you don't want to be like 'oh you're bad because you're a girl'. And you don't want to be the worst and then people would be like 'Oh' (Hannah, Y13).

As her extract exemplifies, this awareness of her gender difference and hypervisibility was experienced as an 'extra pressure', such that she worried that any individual academic 'failings' would be interpreted as due to her gender ('oh you're bad because you're a girl'), thus justifying and reinforcing her gender 'illegitimacy' within the field of physics. Moreover, while the idealised notion of the 'effortlessly clever physicist' may not be restricted solely to female students (as discussed in our other chapter in this volume, discussing Victor's physics trajectory), it was an even more difficult and higher stakes challenge for young women to negotiate.

Reay et al. (2009) describe how working class students' habitus was fashioned and refashioned through their experiences of elite higher education and yet this process did not require or result in a whole sale change, abandonment or 'escape' from valued aspects of their working-class self and family connections. Likewise, we suggest that Hannah did not abandon all performances of femininity and remained intelligible as a young woman (for instance through her appearance, dress and performances of self). However, in line with research conducted with women physicists in higher education (e.g. Ong 2005) who self-consciously 'manage' their femininity, Hannah described 'balancing' aspects of her femininity, for instance, wearing a skirt to school, despite preferring jeans, to counterbalance perceptions of her 'very short' hair. We thus hypothesise that she will continue to have to work hard to regulate and negotiate these balancing acts (so as to maintain intelligibility both in and beyond physics) as she progresses through the physics 'pipeline'.

Consequently, we suggest that Hannah's self-recognition as being 'good at physics' (and hence her ongoing physics trajectory) was precarious. Not only did she need to keep performing and re-performing high attainment so as not to be officially debarred at each educational level, but this performance was also required in order for her to retain symbolic/ representational legitimacy as a physicist and, in particular, to mitigate the illegitimacy of her female body within the field of physics. Thus Hannah faced the difficult challenge of having to manage the contradiction of her feminine body, working hard to maintain her simultaneous intelligibility as a physicist and as a young woman. While her impressive habitus and resources enabled her to navigate a viable position within the field, we would argue that through the maintenance of the notion of the 'effortlessly clever physicist', the field has continued to maintain gender inequality and reproduce the dominance of masculinity. Hence we understand both her reservation that her attainment in physics is not effortless and her 'hypervisibility' as a girl in an Advanced level physics class, as exemplifying the ever-shifting ways through which the field maintains a constant dominant masculinised configuration of physics, such that the subject remains aligned 'naturally' with masculinity, despite the presence of female physicists.

Reay and colleagues draw on Puwar's (2004) notion of 'familiar strangers' to describe working class students in elite Higher Education, who are 'fitting in as learners despite their class difference' (Reay et al. 2009, p. 1115). We suggest that our data also highlight the ways in which the field of physics cultivates the habitus and hexis of students to ensure that only those 'strangers' who are 'familiar enough' are allowed entry and to remain. That is, the field of physics cultivates a particular habitus and hexis through female students' 'minds' and bodies which enables the continued reproduction of the elite nature of the field.

2.7 Discussion and Conclusions – What Is the Future for Young Women in Physics?

In this chapter we employed a feminist Bourdieusian lens to explore what insights it might offer for our understanding of young women's trajectories in, through and out of physics between the ages of 10–19. Our analysis suggests that the field of physics is tightly regulated and strongly orientated to the reproduction of the elite status of the subject. It is arguably a highly efficient and successful field in this respect, given that the eliteness of the field widely recognised and unquestioned and the integral supporting practices and propositions are accepted as doxa. Moreover, like Reay et al. (2009), we found no particular examples of 'painful dislocations' (Baxter and Britton 2001) and no habitus of recalcitrance (Skeggs 2004) among our sample of A level physics girls – even among those who had been expelled/ debarred. Indeed, we found that the field was remarkably effective in cultivating the bodies and minds of young women physicists to fit the needs and demands of the field – enacting a symbolic violence through which young women blame their own abilities when they are debarred and self-exclude from continuing further. We interpret this as attesting to the power and success of physics in performing its inculcation 'job'.

As Skeggs (2004) discusses, subject choices reveal the dialectic of gender and habitus, such that "The transference of femininity from the student to the school subject and back again to the student exemplifies the dialectic of objectification and embodiment, formed via an 'elective affinity' shaping the habitus' (Skeggs 2004: 22). In our chapter, we explored the complicated dialectics involved when young women negotiate an elite subject that is aligned with masculinity – and the intricate negotiations of embodiment and habitus that are involved.

Our data paint a picture in which physics appears to be a risky and challenging option for young women. Indeed, we argue that Davina, Kate and Mienie's choices *not* to pursue the subject post-18 can be seen as rational and strategic, given the intractability of gender inequality within the field which render success harder and more precarious due to the unequal 'rules of the game'. As Beck reminds us, risk adheres inversely to the social structure, such that those in positions of power enjoy fewer risks, whereas those who occupy less prestigious positions experience an ‘unfortunate abundance’ of risk (1992, p. 35). Thus physics remains a more risky option for girls than boys and we can see that Davina, Kate and Mienie may have a better chance of ‘winning’ by pursuing other subjects at university. Indeed, just as elite higher education is a choice that is characterised by ‘conscious deliberation and awareness’ (Reay et al. 2009: 1110) among working class students, so we suggest that for young women, physics is also both a ‘risky’ and ‘dissonant’ choice that needs to be carefully thought through.

The application of our conceptual lens enabled us to identify some key factors and relations which help to explain Hannah’s successful trajectory on to a physics degree. Pre-requisites included the continued performance of high attainment, gender alignment with the field, high levels of physics-specific capital and gender reflexivity. Crucially, Hannah also maintained a self-identification as ‘good at physics’, despite the dominance of the fantasy of the ‘effortlessly clever physicist’ which threatens to undermine her legitimacy and ensure and produce the reproduction of the field’s elite status. While we have not had space in this chapter to explore the intersection of gender with ethnicity and social class, we note here that from our Bourdieusian analytic perspective that, given the pedagogic work that is performed by physics to maintain its elite status, it is not surprising that the only young woman in our sample who ended up pursuing physics at degree level was white and upper middle-class. In this respect, we suggest that the field of physics demands that those young women who are able to continue with the subject are exceptional, privileged and, in Davina’s words quoted earlier, “not like your average person”.

What hope is there for change and greater gender equity within post-compulsory physics? Our data show how a student’s interest, enjoyment, aptitude and passion for the subject are not necessarily sufficient to enable them to pursue it further (Archer et al. 2010). As in the case of Danielle and Davina, even long-held interests and aspirations can be denied and/or ‘cultivated away’ by the field.

Bourdieu argues that pedagogic action – despite entailing symbolic violence – can produce an emancipatory reflexivity:

> … the possibility of an emancipation founded on awareness and knowledge of the conditionings undergone and on the imposition of new conditionings designed durably to counter their effects (Bourdieu 1999a: 340).

While not wishing to dismiss Bourdieu’s uncharacteristic optimism, we suggest a more cautious interpretation in the case of women in physics. While we certainly welcome the cultivation of feminist reflexivity among physics students – and recognise that this may be productive on numerous levels – we argue that without addressing the strict regulation of the field and the underlying lack of value accorded to femininity (and other axes of inequality) both within and beyond physics, the social justice potential remains constrained. Indeed, McRobbie (2004) argue that social change comes about not through the resistance of subordinated groups or through reflexive individualization but through shifts in the conditions of social reproduction. As Adkins (2002, 2004) argues, individualization can bring new social divisions into being – explaining how 'reflexivity concerns not a freedom from gender but is actively reworking the social categories of gender' (Adkins 2004; p. 9).

In particular, we suggest that significant change will only be achieved via transformation of the field itself. Such a change would require addressing the technologies and practices – the pedagogic work – which produces the pedagogic action that reproduces the elite status of physics. Hence, we call for the strict gatekeeping practices around attainment to be dismantled and opened up, not least given the key role these practices play in reproducing the doxa of 'hard' (masculine) physics and the notion of the effortlessly clever physicist.

However, we believe that the likelihood of there being any impetus (or not) for such a change will depend on the continued 'success' of the subject. That is, it is likely that change will only be prompted by necessity. Recent policy concerns about the 'crisis' in physics participation (Saltelli and Funtowicz 2017; Wong 2016) provide a potential point of leverage here. To date, physics has been arguably highly effective in maintaining its elite status by not letting in the 'wrong' people – who might dilute and/or challenge the subject's elitism – and by ensuring that those who do gain entry are socialised into accepting the status quo. However, arguably these restrictive practices have also led to the subject only 'just' surviving in terms of ensuring a sufficient volume of students entering and continuing with the subject to ensure a sufficient economic rate of return – an issue that is being played out against a backdrop of Higher Education expansion (Smithers et al. 2009). Within this neoliberal market logic, (high achieving, middle-class) women have been identified as constituting a potential resource 'pool' of future physicists, which could sustain and sure up the viability of both physics and the wider national knowledge economy (Raelin et al. 2014; The Royal Society 2008). Yet, the individualised approaches and strategies designed to encourage young women to continue with post-compulsory physics which have typically been pursued to date (being those which are most palatable and in keeping with the logic of the field and the reproduction of the subject's elitism) have failed to produce significant changes in the supply of physics graduates (e.g. Murphy and Whitelegg 2006). We thus suggest that it may be an emancipatory reflexivity within the *field* (rather than just within individual young women's habitus) that is required in order to produce emancipatory potential and to substantially improve gender equity – and young women's possibilities – within the subject. In other words, we suggest that the challenge (and potential) will lie in getting the field of physics (and the myriad of powerful actors within this field) to understand the ways in which social reproduction functions in this space – and to then accept a reduction in their previously-enjoyed privilege in order to genuinely redress the effects of inequality and to open up the field to a more diverse demographic of participants.

References

Adkins, L. (2002). Reflexivity and the politics of qualitative research. In T. May (Ed.), *Qualitative research: Issues in international practice*. London: Sage.

Adkins, L. (2004). Introduction: Feminism, Bourdieu and after. *The Sociological Review, 52*, 1–18. https://doi.org/10.1111/j.1467-954X.2005.00521.x.

Archer, L., DeWitt, J., Osborne, J., Dillon, J., Willis, B., & Wong, B. (2010). 'Doing' science versus 'Being' a scientist: Examining 10/11-year-old Schoolchildren's constructions of science through the lens of identity. *Science Education, 94*(4), 617–639.

Archer, L., DeWitt, J., Osborne, J., Dillon, J., Willis, B., & Wong, B. (2012a). Science aspirations and family habitus: How families shape children's engagement and identification with science. *American Education Research Journal, 49*(5), 881–908.

Archer, L., DeWitt, J., Osborne, J., Dillon, J., Willis, B., & Wong, B. (2012b). "Balancing acts": Elementary school girls' negotiations of femininity, achievement, and science. *Science Education, 96*(6), 967–989.

Archer, L., DeWitt, J., Osborne, J., Dillon, J., Willis, B., & Wong, B. (2013). Not girly, not sexy, not glamorous': Primary school girls' and parents' constructions of science aspirations. *Pedagogy, Culture & Society, 21*(1), 171–194.

Archer, L., Moote, J., Francis, B., DeWitt, J., & Yeomans, L. (2017). The 'exceptional' physics/engineering girl: A sociological analysis of longitudinal data from girls aged 10-16 to explore gendered patterns of post-16 participation. *American Educational Research Journal, 54*, 88–126.

Archer, L., Moote, J., & MacLeod, E. (2020). Learning that physics is "not for me": pedagogic work and the cultivation of habitus among Advanced Level physics students. *Journal of the Learning Sciences*, https://doi.org/10.1080/10508406.2019.1707679.

Baker, D., & Leary, R. (1995). Letting girls speak out about science. *Journal of Research in Science Teaching, 32*(1), 3–27.

Baxter, A., & Britton, C. (2001). Risk, identity and change: Becoming a mature student. *International Studies in Sociology of Education, 11*, 87–102.

Beck, U. (1992). *Risk society: Towards a new modernity*. London/New York: Sage.

Blickenstaff, J. C. (2005). Women and science careers: Leaky pipeline or gender filter? *Gender and Education, 17*(4), 369–386.

Bourdieu, P. (1984). *Distinction: A social critique of the judgement of taste*. Cambridge, MA: Harvard University Press.

Bourdieu, P. (1990). *The logic of practice*. Cambridge: Polity Press.

Bourdieu, P. (1999a). Scattered remarks. *European Journal of Social Theory, 2*(3), 334–340.

Bourdieu, P. (1999b). *The weight of the world: Social suffering in contemporary society*. Oxford: Polity.

Bourdieu, P. (2001). *Masculine domination*. Stanford: Stanford University Press.

Bourdieu, P. (2005). *The social structures of the economy*. Cambridge: Polity.

Bourdieu, B., & Passeron, J. C. (1977). *Reproduction in education*. London: Society and Culture.

Bourdieu, P., & Wacquant, L. J. D. (1992). *An invitation to reflexive sociology*. Cambridge, UK: Polity Press.

Calabrese Barton, A., & Brickhouse, N. W. (2006). Engaging girls in science. In C. Skelton, B. Francis, & L. Smulyan (Eds.), *The sage handbook of gender and education* (pp. 221–235). Thousand Oaks: Sage.

Calabrese Barton, A., Tan, E., & Rivet, A. (2008). Creating hybrid spaces for engaging school science among urban middle school girls. *American Educational Research Journal, 45*(1), 68–103.

Carlone, H. B. (2004). The cultural production of science in reform-based physics: Girls' access, participation, and resistance. *Journal of Research in Science Teaching, 41*, 392–414.

Danielsson, A. T. (2012). Exploring woman university physics students "doing gender" and "doing physics". *Gender and Education, 24*(1), 25–39.

Darke, K., Clewell, B., & Sevo, R. (2002). Meeting the challenge: The impact of the National Science Foundation's Program for Women and Girls. *Journal of Women and Minorities in Science and Engineering, 8*, 285–303.
Francis, B., & Skelton, C. (2005). *Reassessing gender and achievement*. London: Routledge.
Francis, B., Archer, L., Moote, J., DeWitt, J., & Yeomans, L. (2017). Femininity, science, and the denigration of the girly girl. *British Journal of Sociology of Education, 38*(8), 1097–1110. https://doi.org/10.1080/01425692.2016.1253455.
Gonsalves, A. (2014). "Physics and the girly girl—There is a contradiction somewhere": Doctoral students' positioning around discourses of gender and competence in physics. *Cultural Studies in Science Education, 9*, 503–521.
Harding, S. (1998). Women, science, and society. *Science, 281*(5383), 1599–1600.
Haussler, P., & Hoffmann, L. (2002). An intervention study to enhance girls' interest, self-concept, and achievement in physics class. *Journal of Research in Science Teaching, 39*(9), 870–888.
Haworth, C. M. A., Dale, P., & Plomin, R. (2008). A twin study into the genetic and environmental influences on academic performance in science in nine-year-old boys and girls. *International Journal of Science Education, 30*, 1003–1025.
Jenkins, R. (2006). *Pierre Bourdieu: Revised Edition*. London: Routledge.
Lawler, S. (2004). Rules of engagement: Habitus, power and resistance. *The Sociological Review, 52*, 110–128. https://doi.org/10.1111/j.1467-954X.2005.00527.x.
McNay, L. (1999). Gender, habitus and the field: Pierre Bourdieu and the limits of reflexivity. *Theory, Culture & Society, 16*(1), 95–117. https://doi.org/10.1177/026327699016001007.
McRobbie, A. (2004). Post-Feminism and popular culture. *Feminist Media Studies, 4*(3), 255–264.
Moi, T. (1991). Appropriating Bourdieu: Feminist theory and Pierre Bourdieu's sociology of culture. *New Literary History, 22*(4), 1017–1049.
Moi, T. (1999). *What is a woman?* Oxford: Oxford University Press.
Mujtaba, T., & Reiss, M. J. (2013). What sort of girl wants to study physics after the age of 16? Findings from a large-scale UK survey. *International Journal of Science Education, 35*(17), 2979–2998. https://doi.org/10.1080/09500693.2012.681076.
Murphy, P., & Whitelegg, E. (2006). *Girls in the physics classroom: A review of the research on the participation of girls in physics*. London: Institute of Physics.
Ong, M. (2005). Body projects of young women of color in physics: Intersections of gender, race, and science. *Social Problems, 52*, 593–617.
Puwar, N. (2004). *Space invaders: Race, gender and bodies out of place*. Oxford: Berg.
Raelin, J. A., Bailey, M. B., Hamann, J., Pendleton, L. K., Reisberg, R., & Whitman, D. L. (2014). The gendered effect of cooperative education, contextual support, and self-efficacy on undergraduate retention. *Journal of Engineering Education, 103*(4), 599–624.
Reay, D., Crozier, G., & Clayton, J. (2009). 'Strangers in paradise'? Working-class students in elite universities. *Sociology, 43*(6), 1103–1121. https://doi.org/10.1177/0038038509345700.
Saltelli, A., & Funtowics, S. (2017). What is science's crisis really about. *Futures, 91*, 5–11.
Skeggs, B. (2004). Exchange, value and affect: Bourdieu and "the self". In L. Adkins & B. Skeggs (Eds.), *Feminism after Bourdieu* (pp. 75–89). Oxford: Blackwell.
Smith, E. (2010a). Do we need more scientists? A long-term view of patterns of participation in UK undergraduate science programmes. *Cambridge Journal of Education, 40*, 281–298.
Smith, E. (2010b). Is there a crisis in school science education in the UK? *Educational Review, 62*(2), 189–202.
Smith, E. (2011). Women into science and engineering? Gendered participation in higher education STEM subjects. *British Educational Research Journal, 37*, 993–1014.
Smithers, A., Robinson, P., & Gatsby. (2009). *Physics participation and policies: Lessons from abroad*. London: Carmichael Press.
The Royal Society. (2008, February). *A higher decree of concern*. Policy Document.
Thomson, D. (2015). *Is A-level physics too hard (and media studies too easy)?* Education Data Lab. Published online on 20th October 2015. Accessed 06/06/2018 at < https://ffteducationdatalab.org.uk/2015/10/is-a-level-physics-too-hard-and-media-studies-too-easy/>

Tracy, C. (2016a). *The problem of inter-subject comparability*. Institute of Physics. Published online on 17 February 2016. Accessed 06/06/2018 at < http://www.iopblog.org/the-problem-of-inter-subject-comparability/>

Tracy, C. (2016b). *Do students choose subjects based on how hard they are graded?* Institute of Physics. Published online on 19 April 2016. Accessed 06/06/2018 at < http://www.iopblog.org/the-effects-of-grading-on-choice/>

Tytler, R., Osborne, J., Williams, G., Tytler, K., & Cripps Clark, J. (2008). *Opening up pathways: Engagement in STEM across the primary–secondary school transition*. Australian Department of Education, Employment and Workplace Relations, Canberra, A.C.T.

Wong, B. (2016). The 'crisis' in science participation. In *Science Education, Career Aspirations and Minority Ethnic Students*. London: Palgrave Macmillan.

Chapter 3
Lighting the Fuse: Cultivating the Masculine Physics Habitus – A Case Study of Victor Aged 10–18

Louise Archer, Julie Moote, and Emily MacLeod

3.1 Introduction

> I mean sort of what got me into Astrophysics is by pure sort of – I wouldn't say by chance or mistake, because I was already along those lines [...] the fuse was already there. What lit it was by mistake (Victor, Year 13 interview, age 17/18)

This chapter attempts to understand how and why it is that (white) middle-class boys, like Victor, are more likely than many other students to end up studying for a physics degree. In particular, we use a Bourdieusian lens to explore how physics identity is shaped by interactions of habitus, capital and field, such that not only are some students more likely than others to see physics as potentially 'for me', but distinctive dispositions are also cultivated and demanded by the field of physics. In short, we examine how physics students 'become themselves' (Webb et al. 2002, p. xii), arguing that, the 'fuse', which Victor refers to (above) as being a precursor to his physics interest, can be understood as constituted through particular configurations of habitus and capital, such that what he calls the 'mistake' that ignites his physics trajectory, when viewed through this lens, is neither random, singular nor unexpected.

L. Archer (✉) · J. Moote · E. MacLeod
UCL Institute of Education, London, UK
e-mail: l.archer@ucl.ac.uk; j.moote@ucl.ac.uk; e.macleod@ucl.ac.uk

A. J. Gonsalves, A. T. Danielsson (eds.), *Physics Education and Gender*, Cultural Studies of Science Education 19,
https://doi.org/10.1007/978-3-030-41933-2_3

3.1.1 A Bourdieusian Lens

Bourdieu's sociological theory provides a compelling framework and a set of valuable analytic tools for understanding the reproduction of social inequalities within society. At the heart of Bourdieu's theory is the idea that interactions of habitus, capital and field produce patterns of privilege and inequality within society.

Habitus refers to the internal framework of socialised, embodied dispositions that a person develops over time. Habitus provides us with 'feel for the game', and a sense of what is normal and desirable for 'people like me' (Bourdieu and Wacquant 1992). It embodies both individual and collective histories – that is, the habitus is not purely individualistic but is profoundly social and collective (or as Bourdieu terms it, "the subject is not the instantaneous ego of a sort of singular cogito, but the individual trace of an entire collective history", Bourdieu 1990: 91). The habitus is both shaped by experience and, in turn, shapes our views and interactions with the world. In Bourdieu's words, it can be understood as a 'system of structured, structuring dispositions' (1990, p. 52), meaning that it is both a product of socialisation (i.e. it is structured) and provides a guiding framework for how we experience, interpret and interact with the social world (i.e. it is structuring).

Capital refers to cultural, social, economic and symbolic resources (or 'accumulated labour', as Bourdieu terms it, Bourdieu 1986, p. 241) that a person may possess and accrue. In this respect, capital is the 'hand' you can play in the game. Capital exists in relationship with habitus and field. For instance, the value of capital is determined by the field within which is operates ('Capital does not exist and function except in relation to a field', Bourdieu and Wacquant 1992, p. 101). Within any given field, the most powerful forms of capital will be those whose value can be most readily converted into symbolic forms that match the requirements of the field. Hence the field determines value of capital but the accumulation of capital also, in turn, shapes the field.

Field refers to socially and historically constructed socio-spatial arenas that are constituted through differential power relations. Field thus represents a 'space of positions and position-taking' (Bourdieu 1993, p. 30), which Bourdieu imagines as a 'force field' that constitutes the 'rules' of the game:

> A field is a structured social space, a field of forces, a force field. It contains people who dominate and others who are dominated (Bourdieu 2010, p. 37)

The relationship between habitus, capital and field will shape a person's experience within that field. For instance, where there is a strong alignment, students experience education as a 'fish in water' (Bourdieu and Wacquant 1992, p. 127). Each field (and sub-field) has its own logic of practice. Hence, as we have argued elsewhere (Archer et al. 2017; Archer et al., 2020), we suggest that school physics can be understood as a field, with its own logic of practice, containing differently positioned social actors (e.g. teachers, students) who deploy and compete for a range of (cultural, social and symbolic) resources.

While Bourdieu's theory has been extensively and successfully applied to understanding the reproduction of social class inequalities (notably through education), opinion has been more divided regarding its application to gender, given that 'Bourdieu's social theory had relatively little to say about women or gender' (Adkins 2004, p. 3s). However, many feminists have found Bourdieu's conceptual tools both useful and resonant with a feminist approach, particularly: his constructivist structuralism and undercutting of the essentialist/non-essentialist divide (Moi 1991); his foregrounding of embodiment (as core to the habitus); the notion of power and social position taking as constitutive of field; the concept of symbolic power/violence; and his non-Cartesian 'notion of the subject as not simply engaged with the world, but *in* the world' (Adkins 2004, p. 10).

Towards the end of his career, Bourdieu did attempt to explicitly engage with gender, as exemplified by his book *Masculine Domination (La Domination Masculine)*, based on an anthropological study of the Kabyle society. However, attention has been drawn to some key limitations of his framework for feminist theory. For instance, because his conceptualisation of gender was based on his study of a comparatively less complex and undifferentiated tribal society, his work has been critiqued for not extending to contemporary, highly complex societies in which there are diverse understandings and performances of masculinity and femininity (that is, it assumes an 'understanding of gender relations in contemporary, differentiated, heterodox societies on an outdated anthropological study of an undifferentiated society', Adkins 2004, p. 16). But also more generally, it is widely recognised that as a consequence of his focus on social reproduction, Bourdieusian theory has difficulty explaining social change. Further critiques have also been levelled at his assumption that the habitus is driven by capital accumulation. For instance, Reay's 2004 analysis of mothers highlights how emotional capital can be expended for the benefit of investing in others, to the detriment of the self. Yet, such reservations notwithstanding, his work remains a productive resource for feminist theory and research, particularly, as exemplified by the contributors to a 2004 special issue on *Feminism and Bourdieu*, which showcased how feminists are reworking, redefining, extending and transforming his work in order to better understand contemporary gender inequalities in western societies (see Adkins 2004). For instance, as McLeod argues, Bourdieu's concept of field is rich and can be usefully extended and applied to theorise how dominant notions of gender (both how it is expressed and experienced) change and evolve across time and context. That is:

> how gender identities and relations are changing or being re-articulated in new but familiar ways. In such a view, a more uneven and less seamless relation between habitus and field is possible, and this offers scope for feminist analysis of change and continuity (McLeod 2005, p. 12)

Moreover, although Bourdieu is grounded in structuralism, as McLeod discusses, his 'denaturalizing' of gender and his view of gender as a socio-historical and arbitrary 'social construction' (Bourdieu 2001, p. 3) fits with feminist post-structural theorizations of gender. As McLeod further explains, Bourdieu's foregrounding of the embodied nature of habitus (hexis) aligns well with much feminist gender theory and concerns with embodiment. However, as McNay (1999) argues, Bourdieusian theory falls down in its assumption of unitary, rather than multiple, gender subjectivity:

> [Bourdieu] significantly underestimates the ambiguities and dissonances that exist in the way that men and women occupy masculine and feminine subject positions … masculine and feminine identities are not unified configurations but a series of uneasily sutured, potentially conflictual subject positions. (McNay 1999, pp. 107–108)

Hence in this chapter, we attempt a feminist application of Bourdieusian theory, exploring what this lens can offer to our understandings of the intractable, gendered (masculine) nature of post-compulsory physics participation – focusing particularly on how some students (notably middle-class boys, especially from white or South Asian communities) are more likely to be socialised into seeing physics as potentially 'for me', including the cultivation of particular (masculine) dispositions and the requirement of particular capital.

3.2 Methods/Data Set

Data drawn on in this chapter were collected as part of the Aspires/Aspires2 project, a ten year, mixed methods study of children's science and career aspirations from age 10–19 funded by the UK Economic and Social Research Council. The dataset comprises large-scale national surveys of a cohort of students as they progress through primary and secondary schooling, combined with in-depth longitudinal interviews with a subset of young people and their parents.

The survey data referred to in this chapter relates to the wave conducted with students at age 17/18 in England. The questionnaire was revised, validated and piloted with 200 students before being administered to a national sample of c.8000 17/18-year-old students (Year 13). Following data cleansing (which involved removal of duplicate or incomplete responses, as well as any participants with missing ethnicity and gender data), 7326 students remained in the sample for analysis. Students from 265 schools completed the survey (237 state-maintained schools; 28 independent schools), which was roughly proportional to the overall national distribution of schools in England by region, school type, attainment and free school meals provision (as a measure of socioeconomic status).

Longitudinal interviews were conducted with 61 young people (26 male, 34 female, 1 other) and 65 of their parents (50 female, 15 male). All these participants had been previously tracked since students were aged 10/11 (Year 6/Y6), with repeat interviews at age 12/13 (Year 8), 13/14 (Year 9), 15/16 (Year 11) and age 17/18 (Year 13). Interviewees came from a broad range of socioeconomic classes and ethnic backgrounds (43 White British [25 female, 17 male, 1 other], 3 Black [2 female, 1 male], 5 White European [3 female, 2 male], 6 Asian [2 female, 4 male],

4 Mixed [3 female, 1 male]). Interviewees had been recruited to take part in a study of their science and career attitudes and aspirations. Interview topics covered a broad range of areas, including general (non-science specific) areas such as general interests, pastimes, constructions of gender identity, peer relations, their family, aspirations and future educational and occupational plans. Science and STEM-specific areas of questioning included views on science, technology, engineering and maths, views on school science and reasons for post-compulsory science (non) participation. Interviews were conducted by members of the research team, who identified as White, middle-class women from British (n = 2) or North American backgrounds (n = 2). Participants were invited to choose a pseudonym to maintain anonymity in any reporting. Therefore the pseudonyms cited in this paper reflect the personal choices of the interviewees. In this chapter, we focus particularly on the case of Victor, the only boy in our interview sample to have entered onto a physics degree course when students were followed up in the Autumn after they had completed school at age 18.[1]

Quantitative analyses of the survey data were conducted by the second author using SPSS (including series of one-way ANOVAs with post-hoc tests to explore group differences across a range of background factors and for categorical variables, descriptive cross-tabulation (chi-squared) analyses to identify basic relationships, with additional adjusted residual analyses performed, see Moote et al. (2019) for further justification of this procedure). Qualitative data were initially coded by the research team using the NVivo software package. In the first sweep, coding was used to identify top level themes such as 'aspirations', 'views of science', 'home and family', and so on. The lead author then subjected the data to a more theoretically-driven analysis in line with the project's conceptual framework, to identify instances of habitus, capital, field, hexis, pedagogic work and masculinity. Relationships between these concepts were then explored and developed through an iterative moving back and forth between coded data and theory/literature to produce a further set of analyses, in which notions of 'cleverness' and physics masculinity were identified and re-applied to the data (to identify further instances).

3.3 The Distinctive Physics Habitus

Across the project as a whole, data collected from the surveys and interviews painted a common picture, in which the vast majority of students (irrespective of their post-16 routes) viewed physics as being not only 'hard' but also the 'most difficult' science (see also Archer et al. 2017, 2020). Statistical analysis of the

[1] One girl, Hannah, had also entered a physics degree course, a number of students had entered STEM degree or engineering courses and a small number of other students were in the process of taking a 'gap year', some of which included re-applying for physics degree courses.

survey data suggested that students who took Physics A level (an Advanced Level highly academic post-compulsory course, typically studied at age 16–18) expressed a distinctive set of views compared to both other science students (that is, those taking biology and/or chemistry, but not physics) and compared to the wider set of all students. For instance, as discussed in Archer et al., (2020), physics A level students were significantly more likely than other students to feel confident in their mathematical abilities and tended to express the most positive views of science. They were also significantly more likely to express stereotypical views of scientists as being 'odd', 'geeky', 'male' (with chi-squared tests showing large effect sizes[2]).

We interpret these patterns as potentially marking out the contours of a 'physics habitus' – that is, a configuration of socialised dispositions that are structured and cultivated by the field of (school) physics and which, in turn, structure young people's views and experiences of the subject and shape the extent to which they feel it is 'for me', or not. We now seek to add richness and depth to the bones of this statistical picture using a case study of Victor, to try to better understand how these distinctive patterns emerge.

3.4 Introducing Victor

Victor is a white British middle-class boy who lives in the South East of England. Since we first met him at primary school, Victor had aspired to a science or STEM-related career. For instance, at the age of 10 he told us "I'd like to be a scientist, an inventor, and an engineer" and, when asked how long he thought he would continue studying science, he asserted "a long time […] I'm going to do that in university". Between the ages of 12 and 14 he wanted to be a science teacher and by the age of 15 he had decided to become an astrophysicist – an aspiration that he continued to hold into university. Victor lived with his mother (a physiotherapist), father (an engineer and master craftsman) and younger brother. He attended his local state primary and secondary school until at age 18, having completed A levels in physics, mathematics and economics, he began an astrophysics degree at a post-1992 university in the South East of England.

[2] Scientists as odd: $x^2(42227) = 82.655$, $p < .001$, Cramer's V = .193, indicating a very large effect; Scientists as geeky: $x^2(42227) = 38.279$, $p < .001$, Cramer's V = .131, indicating a large effect size; Scientists as male: $x^2(42228) = 101.154$, $p < .001$, Cramer's V = .213, indicating a very large effect).

3.5 Victor's 'Physics Habitus' and Dispositions

Victor's physics habitus was consistent with the patterns observed in the survey data, in that he was confident in his mathematical abilities, expressed positive views of science and tended to see science as male-dominated. Analysis of his interview data also allowed for further fleshing out of his physics habitus, which as discussed next, included: a high level of STEM literacy and frequent 'talking science'; a particular interest and fascination with physics; and intellectual curiosity in the way the world works. As discussed next, all of these aspects were evident not only in his academic engagement but also suffused Victor's leisure time.

Across his interviews, Victor exhibited considerable STEM literacy from an early age and regularly 'talked science' (Lemke 1990) in the interviews. From the age of 10 onwards, he conveyed a knowledge of scientists (e.g. Stephen Hawking, Einstein), inventors and engineers (e.g. Benz) and displayed a mastery of scientific terminology. Indeed, all his interview transcripts were peppered with examples of his use of scientific terminology and concepts. For instance, when Victor was ten, his mother, Sam, reflected on his everyday use of scientific concepts and language, which she termed 'the right kind of words':

> I think the other day someone said to him 'What do you use a mirror for?' and he said 'It reflects light' – after someone else had answered. And I thought 'Oh right, you're just using the right kind of words' (Sam, Year 6 interview).

Victor's physics habitus was characterised by a general interest in science and a particularly keen passion for physics. As he reflected in his Year 9 interview, "I'm always thinking about Science in my head and I think that's a good thing". He also recognised in this interview that his interest in science was continuously increasing (or 'excelling' in his words). For instance, when asked how he thought his interest in science had changed since he was younger, Victor replied:

> I think it's sort of excelled a bit and I've more of ... I think more of my time is spent thinking about Science than thinking about other things. Every now and then when I start thinking about something, it somehow always leads towards the plan that I have for doing something with hydroelectricity but ... yeah. [...]... I'm always drawing or sketching something along the lines of Science (Year 9)

It was in his Year 11 interview that Victor's interest in physics came to the fore: "I've seen what sort of all of the areas are in-depth and Physics is the one that fascinates me the most". Victor did not see physics as just a school subject, rather it was a source of personal interest and identity, both in and out of school. For instance, when asked about his out of school interests, Victor replied:

> At the moment I'm into playing squash with my dad on Saturdays. Um, I also quite like computer gaming as well, but apart from that my main interest is Astrophysics. I just kind of like space quite a lot (Year 11).

This interest continued to increase through A level into a real passion for physics: "Physics is great, I love Physics. Er, I love the concept of Physics" (Victor, Year 13). Victor's science/physics habitus was also underpinned by an intellectual curiosity and interest in how the world works. As he recounted in his first interview, age 10:

> Well I just think it's fascinating the way things work. I've got this book on the way things work and it was very big, and then they just had things that were invented like in the olden days, but we still use today (Year 6).

Indeed, from when we first met him, Victor had defined himself through his science activities and interests:

> **Interviewer**: So we'll start off quite basic – can you tell me a little bit about yourself?
> **Victor**: Um, well I like doing things to do with science. I like making things, making things move, and then I like making things go up in the air. [I: Okay] I tried that with baking soda and vinegar – I tried it with my brother and it was so much fun. Um, I also like animals and wildlife. (Year 6)

By the age of 10, he had created a book, or journal, in which he listed all his ideas and plans for inventions:

> I've put it down all in a book called 'Impossible Things'.. I've got a time vortex, a cloning thing, robots that do like everything, and loads of other things (Year 6).

Across the interviews he also described setting himself personal research projects, which reflected his long-held ambition to become a scientist, as the following extract from his Year 6 and Year 9 interviews exemplify:

> **Victor**: I want to be a scientist because … well science is like … when you think about it, science is just amazing. I've given a question to myself, and I'm trying to do a little bit of research on it – it's what creates wind.
> **Int**: Mm. And what have you done so far?
> **Victor**: Well so far I'm not too far through, and all I know is that wind is like a pushing force.
> **Int**: And how do you plan to find out more about what creates wind?
> **Victor**: I can use the internet and then I can like use my … computer and friends that want to be scientists. (Year 6)

> **Victor**: I'm interested in creating things at the moment, inventions, I'm thinking of doing something, I don't know if it will work out, but um … yeah […] I've been thinking of … along the lines of hydroelectricity, I'm thinking of the generalities and everything and I'm not entirely sure, but it's along those lines. (Year 9).

To understand what produced this distinctive habitus, we now apply a Bourdieusian lens, exploring how interactions of capital, habitus and field, along with specific alignments of his habitus and capital with particular notions of masculinity and 'cleverness' that are demanded and normalised within physics, combine to possibilise Victor and his trajectory to being/becoming a physics student.

3.6 Victor's Science-Related Family Habitus and Capital

As has been widely noted in the literature, the middle-classes tend to be highly successful at combining economic, cultural, and social capital to produce academic achievement (e.g. Dika and Singh 2002). Through the process of socialisation and the translation of capital, these families are able to cultivate particular values, attitudes, expectations, and behaviors in their children that enable them to succeed educationally (e.g. Israel et al. 2001; Martin 2009; Perna and Titus 2005; Sandefur et al. 2006), including through private schooling, private tuition, and extensive enrichment activities (e.g. Vincent and Ball 2007). However, as we have noted previously, in the case of promoting successful engagement and trajectories in science, it is a family's possession and deployment of specific science-related forms of cultural and social capital (and the deployment of economic capital to support the accrual of science capital, Archer et al. 2015 – I.e. science-related forms of cultural and social capital, as embodied in the habitus) that is particularly influential (DeWitt et al. 2016). Indeed, our research has found that young people with high levels of science capital are significantly more likely to aspire to continue with science post-16 and to espouse a 'science identity' (e.g. Archer et al. 2015.)

Victor grew up in a home rich in STEM-related capital. His parents were STEM enthusiasts (as his mother, Sam, explained in the first interview, "I'm hugely, hugely interested in science") and had STEM-related jobs, in physiotherapy and engineering respectively. Victor's grandfathers on both sides were also engineers, which his mother interpreted as an essentialised, or almost inherited, family connection, using the terminology of it being 'genetic': "Well my husband's an engineer, his father was and my dad was, so I think it's, you know, quite genetic with [Victor]").

Victor enjoyed considerable STEM-related social capital, not just through his family members but also through their wider social networks. For instance, from the age of 10, Victor told us about the number engineers he knows, naming several of his father's friends. He also benefitted from social capital in the form of like-minded peers. For instance, his early primary friendships enabled him to enjoy talking about science with peers and to have a rich, playful and socially rewarding engagement with science which also provided Victor with new science capital and knowledge. For instance, as exemplified in the following extract, Victor's friend, Finch, liked science and had introduced him to nuclear physics and the periodic table. The boys also used to enjoy science-themed imaginative play together:

> Well I think my friends are [into science], cos I think it was nuclear physics that [Finch] liked and ... well I talk to him a lot and then he just goes off talking about hydrogen, oxygen, the other one that he ... he did the whole of this sort of table thing – I can't remember its name but ... I think it was the periodio- ... no ... can't remember. // Well [Finch] has the same desire to make things that go zap. We're trying ... in our imagination games there's something called the zappy and it can teleport you, and you can also use it as a laser. [Int: Okay.] We tried making it at his house when I went round there, and then I tried to make one of his machines work, but then it did an overdrive and one of the batteries leaked. (Victor, Year 6).

Victor continued to have 'science-y' friends through secondary school – which again provided further spaces and opportunities to develop a science habitus. As the following quote exemplifies, these relationships also provided cultural capital and, as Victor suggests, made him first aware of the field of astrophysics:

> I think a friend just mentioned the word Astrophysics and I was thinking I like Physics and I like Astronomy, because I was really keen on Astronomy sort of thinking I really like sort of the space itself and things like that and everything in the universe, but then I didn't really want to just study the stars. I thought that's been done quite a lot in history. I want to do something along the lines of how everything works and then my friend said Astrophysics and sort of by chance that's exactly what I want to do (Victor, Year 11).

As we have previously found to be common among 'science families' (those with high levels of science capital, Archer et al. 2012a, b), Victor's childhood was also characterised by frequent informal engagement with science, through regular consumption of science-related television and media and through his many home science resources, such as science kits ("yeah I've got one of those kits where you just mix things together and make different things. Once I made a big explosion", Year 6). For instance, at age ten Victor told us about how he liked making moveable vehicles and was planning on building a helicopter. He also enjoyed reading science books, with a notable one being astrophysics-related ("it was called 'My First Guide to the Universe' and I learnt about how things were affected by gravity"). His mother concurred, and described the motorised cars and rockets that he had built and how he liked making explosions, using science kits and "he was trying to make an electrical circuit recently, you know make a bulb light up" (Sam, Year 6).

As our previous research has shown, families with higher levels of science-related resources (capital), tend to actively promote, develop and sustain their children's science interest and aspirations, through the foregrounding of science within everyday family life, for instance, by providing science kits, watching science TV together, discussing science in everyday conversations, going to science museums, and so on (Archer et al. 2012a, b). In this respect, Victor's family can be read as a typical example of such a 'science family'.

Victor also recounted trips to the science museum, which had made an enduring and visceral impression on him, explaining how he 'can still feel' the effects of a particular exhibit when he remembers it:

> It was a long time ago. I can remember playing a game and then there was this long pillar in the middle and it was like full of static, and then I touched it and then the static just came down the line and you're like 'Ahh' and it tickled me. Yeah I can still feel it in myself. (Year 6)

He also described being inspired by the famous physicists that he learned about through his consumption of science media. For instance, at age 13/14 he explained:

> Well Science, I got into because, you know, people like Stephen Hawkins and other people like Einstein, Newton, I thought there was something there because like the science of the world is really complicated […] So I thought like I could contribute to doing that or … um it really came about when I was doing Science because I was fascinated by the fact that if you had this and you mixed it with this, it becomes something else which can do this. I

> found that really interesting that ... just what ... what it ... what nature can do by itself I thought was really interesting (Victor, Year 9).

Over the years, Victor continued to do a substantial amount of STEM-related activities in his leisure time, although these shifted more towards science-related computer games. For instance, at age 15/16 he described his love of a space-related game, which he claimed has sparked an interest in astrophysics:

> There's one called Kerbal Space Program. It's a simulator of space and that's what's kind of got me into the whole Astrophysics sort of thing [...] you get sort of command modules and things and fuel tanks and engines and different types of things and you have to sort of teach yourself how to stage things properly, how to enter an orbit, (inaudible 00:16:34), things like that and it's got a whole solar system, so you can go and visit loads of planets. It's all real, sort of real time in game, so all the planets move round [...] it's accurate sort of in proportion to the solar system, but it's got like I think the atmosphere's thinner, so you can get out into orbit a lot easier (Year 11).

His out-of-school interest and engagement was further supported and reinforced through his social capital, for instance, in his Year 11 interview, Victor described going out stargazing with a friend of his father.

Victor's home capital was operationalised and realised through his parents' practice of concerted cultivation (Lareau 2003), which both Victor and Sam recognise as being consistently enacted over time. For instance, in the Year 6 interview, Sam described her 'hands on' parenting style:

> Yeah I feel I am quite involved as a parent, I do his reading with him and his homework. Yeah ... no I feel very involved actually (Sam, Year 6)

Similarly, she described how her close involvement was still evident when Victor was aged 18, exemplified by the considerable effort that she put in to researching and helping Victor to explore post-18 routes, such as higher apprenticeships and degree courses. Sam was highly industrious (and the main driver in) seeking information, requesting help from the school and extensively researching different options:

> Perhaps I'm just being a helicopter parent and wanting too much, I don't know. But I just feel that the work that has been done to find everything out has just been me [laughs] (Sam, Year 13).

As Sam explained, "that's how we look at things, so you know if that door closes, it means that we've got to keep looking, pushing other doors". Bourdieu (1986) describes capital as accumulated labour and argues that the embodiment of capital and its incorporation into the habitus, such as through tastes, dispositions, and so on, is gradual and takes time. We interpret Sam's industriousness as exemplifying the considerable labour that is required to convert capital into the habitus and to realise it as social advantage – it is not an automatic or passive process. In other words, producing one's child as a viable, successful scientist requires considerable time, resource, strategy and 'work' to ensure the 'correct' translation of capital into social advantage. Bourdieu argues that the conversion and transformation of capital requires labour, particularly time and effort by mothers, to generate cultural capital. While we may take some issue with Bourdieu's somewhat gender stereotypical assumption that mothers are necessarily the prime socialising agent, we find that his description of this process (as "diffuse, continuous transmission within the family [that] escapes observation and control") fits well with our data in the case of Victor, as indeed does Bourdieu's conclusion that the purpose of this process is so that "the educational system seems to award its honors solely to natural qualities". (Bourdieu 1986, pp. 25–26).

3.6.1 School

Beyond his home capital, Victor also described deriving resources and support from particular school teachers, who provided another 'spark' for his interest and aspiration:

> Our old Physics teacher [name] was really good at teaching. He was also quite interested in space and that sort of, I could talk to him about things that were going on and that sort of sparked my interest (Year 11)

His mother, Sam, also recognised the value of this teacher for supporting Victor's trajectory:

> They had a new Physics teacher towards the end of last year who is an astrophysicist [...] Thank God for that (Year 13)

However, Sam also described how this capital was variable, in that not all of Victor's physics teachers had been supportive or effective. Indeed, she felt that he had "done well for a state school student", given that many state schools in England (like Victor's) experience particularly acute issues with the recruitment and retention of physics teachers (e.g. Osborne and Dillon 2008). Sam described his science teachers as having been "either superb or appalling", explaining: "He's had some superb teachers, he's got some at the moment … he's had some absolute stinkers … but that is because schools can't get science teachers".

The family negotiated the challenges and risks posed by the perceived variable quality of school science teaching by mobilising their economic and social capital to provide Victor with private tutoring, to support his attainment. As Bourdieu argues, the habitus can be strategic and spontaneous, enabling his family to be responsive to new situations (such as issues with teaching and threats to his attainment), albeit within the limits of what is 'thinkable' (as unconsciously driven and guided by the socialised habitus).

> 'the habitus, like every 'art of inventing', is what makes it possible to produce an infinite number of practices that are relatively unpredictable (like the corresponding situations) but also limited in their diversity. In short, being the product of a particular class of objective regularities, the habitus tends to generate all the 'reasonable', 'common sense' behaviours (and only these) which are possible within the limits of these regularities.. (Bourdieu 1990, p. 55)

In other words, his family's strategic deployment of capital can be understood as both a conscious and unconscious function of the habitus, guided by the imperative of the social reproduction of privilege (i.e. to try to ensure class advantage through the production of educational 'success').

3.7 The 'Naturalisation' and Embodiment of Victor's Science/Physics Identity/Habitus

Beyond the mobilisation of capital, we also wanted to explore some of the mechanisms through which a specific science/astrophysics identity was cultivated and embodied within Victor's habitus. Our reading of the data identified a cluster of narratives that seemed to be influential and important to this process and which 'possibilise' (Butler 1990) and mark Victor as a legitimate science subject. For this aspect of the analysis, we combine our Bourdieusian lens with a feminist poststructuralist lens.

Our feminist reading of Bourdieu recognises a performative aspect of habitus, in which identity performances are produced through and with the 'embodied history' of the habitus (which is turn is produced through a dialectical relation with the field). That is, we posit that the identity discourses which a person has to draw on will be shaped by their history, body, capital and interactions with the field. As discussed above, we conceptualise the habitus as strategic, performative and productive – albeit within the limits of its conditions of production. In this respect, we consider the identity narratives produced by Victor, his family and wider others (e.g. school teachers and peers) as being both structured by, structuring and productive of, the habitus.

Over the years, Victor articulated a set of coherent identity narratives and performances in which he produced himself as a 'science person'. For instance, he often described himself as 'always thinking about science'. He also presented himself as a dedicated science student ("Well, science I put in 99% [effort] like all the time. Other subjects I'm like 80, 90%") and felt he was achieving well and had a 'natural' aptitude for the subject ('always getting the stuff that we need to do down straight away");

> **Interviewer**: And how do you think you're doing in science at the moment?
> **Victor**: I think I'm doing very well.
> **Interviewer**: Mm. And why do you think that?
> **Victor**: Well it's because every time we do Science I'm like always getting the stuff that we need to do down straight away.

Importantly, Victor's self-identification as a 'science person' was also recognised and legitimated by others. In particular, across all the interviews, his mother, Sam, articulated a strong and consistent discourse that naturalised and essentialied Victor's science identity as something that is in his 'genetics', 'bones' and 'brain'. For instance, as quoted earlier in the chapter, Sam described how having various engineers in the family (two grandfathers and his own father) rendered engineering 'quite genetic' ('natural') for Victor. She also made references to engineering being 'in his bones', 'brain' (e.g. "[Victor] has got kind of an engineering brain", Year 6) and as being a 'natural' aptitude that he was 'born with':

> Yeah Science ... well in Year 3 he got like 97% in his Science or something – that's something that seems again to come easier to him I think – he's quite fascinated. [...] He was doing something on magnets the other day and he kind of just understood about forces

> of repulsion … yeah so that really delights me actually when I see him doing that kind of thing, because I think 'Oh right, that seems to be where your natural ability lies' (Sam, Year 6).

> [Victor] always wanted something on the moon or satellites or how things work – all this kind thing, he's always just had that kind of brain, taking things apart […] Sometimes you're just born with those kind of … yeah. And I think … yeah I just think that's something that he was just sort of born with, just being very interested in how things work (Sam, Year 13).

In their later interviews, Victor and his mother both recalled examples from earlier in his life which they explained as demonstrating a 'natural' and enduring interest in space. This became a familiar and well-rehearsed family identity narrative over the years:

> [I remember] when we went to the library to get sort of space books and my mum told me the other day 'you always used to go and get like a book on Astronomy and then tell me stuff about it' (Victor, Year 11).

Bourdieu suggests that the habitus is 'embodied history, internalised as a second nature and so forgotten as history' and is thus 'the active presence of the whole past of which it is the product' (Bourdieu 1990, p.56). We suggest that Victor and Sam's talk can thus be read as active performances which both remember and constitute (as identity practices in their own right) this embodied history. That is, we suggest that such family narratives (through their repeated re-telling over time) play a part in producing, embodying and cementing a particular science/astrophysics identity within the habitus. Moreover, they are part of the dialectic through which the workings of capital and field can be 'forgotten' or erased, and the resulting habitus produced as 'natural' (and therefore both legitimate and authentic).

The power of these family narratives (regarding Victor's 'natural' science identity) were further reinforced and legitimated (and hence in turn, cemented further in the habitus) by the wider recognition that Victor also received from others, notably peers and teachers. Although we did not interview Victor's teachers and friends, he and his mother both described numerous examples of such recognition of his alignment with STEM identity and a STEM trajectory. For instance, in her Year 13 interview, Sam described how in the end of school student awards, Victor had been voted the 'most likely person to invent something'. As Carlone and Johnson (2007) discuss, it is the combination of both self-recognition and recognition by others that is key to the achievement of a (successful) science identity – or as we would see it, as key to the performance and cultivation of an enduring science habitus.

3.8 Masculinity and Victor's Physics Identity

So far, we have argued that the combination of habitus, capital and field made Victor's journey to degree level physics less surprising or 'chance' than his opening quote suggests. Now we identify and consider two further aspects which are important parts of the dialectic for understanding Victor and his astrophysics trajectory – masculinity and cleverness.

As noted in the wider literature, physics is a subject/discipline that is strongly aligned with masculinity (e.g. Danielsson 2012; Gonsalves 2014). Our analysis suggests that it is not the mere fact of possessing a body marked as male that is key, per se, to development of a physics identity and trajectory. Rather, Victor's case suggests that it is his embodiment and enactment of a particular performance of masculinity that is key to the possibilising of his physics identity and trajectory.

Across the eight years that we interviewed Victor, he consistently described his own masculinity (and that of his small group of close friends) as being 'different' to most other boys, although he could not always articulate what this difference was. As he put it in his Year 9 interview, "there's a definite difference between me and say, other ... other boys. And I know there's a difference there but I don't know what it is". Sam concurred:

> He's always been his own boy, he's never sort of followed the ... you know just when he was at junior school you know everyone used to turn up to a disco in a football kit, and he'd go in his cowboy suit kind of thing. [laughs] He doesn't really beat to the drum of other people, he's just always found his own path, and he's never been ... you know if he doesn't want to do something that's it, he's not going not going to do it. But if he's going to do it, he'll do it with all his heart kind of thing. (Sam, Year 13).

When probed, Victor did identify some particular dimensions of difference. For instance, he explained how, unlike many of their male peers, he and his friends were not 'laddish' (Francis 1999) – for instance, Victor behaved well at school and did not like getting into trouble:

> I know I'm going to be a bit different because, first of all ... well, all my other friends do sort of things that I wouldn't necessarily do, that makes me quite different from them. [...] I sort of act different to them because since I don't like getting in trouble, since I don't like getting in trouble for a laugh, I think that's what makes me different from other students (Victor, Year 8).

Similarly, Sam described Victor as: "quite a sensitive boy, sort of a bit scared of getting told off and that kind of thing. He does try his best" (Year 6). More particularly, we read Victor as embodying/performing aspects of what might be termed 'geeky masculinity' – namely being academic, into computer games, being a bit 'square' and 'conservative' rather than into macho masculinity and popular culture and somewhat introverted (see Mendick and Francis 2012). He was also aware that some peers positioned him as geeky. For instance, Victor described feeling distinctive and 'different' from other boys by dint of his high attainment and interest in learning:

> Well, I'd say I'm a bit different. Uh, one of my friends said, in Year 5 [...] everything changed ... and I'm ... I'm in set one, I realised that I was ... I realised that I was the only boy in my ... in our form that was in the top set (Victor, Year 9)

As Francis (2000) discusses, popular laddishness is organised around the performance of 'anti-education' sentiments and derides bookish and studious behaviour as 'unmanly'. Victor himself conveyed some awareness that he might be positioned by some (laddish) peers as 'geeky' in this respect. For example, he recounted an instance when he had been mocked by another boy:

> [Name of boy] always … when I say something, he always says something. He just goes into some really weird voice and then pretends to push up some broken glasses. And then he just says 'Oh, I think about stuff' in his really random voice in a sort of weird way (Victor, Year 8).

Victor also eschewed key aspects of laddish popular masculinity, such as a pre-occupation with football and drinking. Instead, Victor espoused a slightly more conservative or 'square' masculinity.

> He plays guitar in church. He helps out with the young people's club on Friday at church for the little children. […] He sees his girlfriend […] um … and he's going to play Dungeons and Dragons with his friends tonight, that's another thing he likes doing (Sam, Year 13)

His performance of masculinity was strongly influenced by his Christian faith, which became more pronounced and important to Victor as he got older. In contrast to extroverted performances of laddish masculinity, Victor felt different from other boys on account of being, as he termed it, more 'held back' and 'not really outspoken', a trait which he felt he shared with his father:

> Different in the fact that I'm a bit more sort of held back, conservative type of thing, not really outspoken. I've probably picked that up from my dad if I'm going to be honest, because he's quite, it's not that he's anti-social. It's just that he gets frustrated at things that aren't going … like if he knows that something can be better (Victor, Year 11)

As Sam put it:

> He's quite quiet on the whole, but has quite strong opinions on things if you ask him. [Victor] is the kind of child who's happiest when he's building Lego, making things. Or his other passion is making comics. Or just … he's very happy to be left to his own … to get on with his own thing – he doesn't necessarily need a lot of company (Sam, Year 6).

In his teenage years, Victor and his close friends became avid fans of the computer game Minecraft:

> we have in common the fact that we all like computer games, the thing is, we all are interested in one specific computer game [Minecraft] (Victor, Year 8).

His passion for computer games continued as he grew older, as discussed previously, such as his love of space-themed games.

As has been noted in the literature, physics is popularly associated with notions of masculinity. For instance, notions of physics 'natural ability' tend to be strongly dominantly aligned with masculinity (e.g. Carlone 2003). In particular, physics is dominantly aligned with 'geeky' (highly intelligent, socially reserved, non-laddish and video-gaming) performances of masculinity (e.g. see Danielsson 2012; Gonsalves 2014; Ong 2005), as epitomised by the character of Sheldon in the US sitcom TV series *The Big Bang Theory* (see Mendick 2016). We thus read Victor's performances of masculinity as strongly aligning with 'geeky' physics masculinity, and specifically the masculinised notion of the 'physics genius', which supports the possibilising of his science/astrophysics habitus. Extrapolating from Bourdieu, we suggest that Victor's successful physics trajectory is not only realised through his deployment of capital but also the cultivation of the 'right kind' of masculine habitus and *hexis* (the expression of habitus through the body), which has incorporated the predominant form of academic, socially reserved, 'geeky' masculinity that is aligned with physics and which, in turn, structures Victor's perception of what is possible and desirable for boys like him. In this respect, we read the interplay between habitus, capital and the fields of physics and masculinity as producing in Victor:

> 'a socialised body: a structured body, a body which has incorporated the immanent structures of a world or of a particular sector of that world – a field – and which structures the perception of that world as well as action in that world (Bourdieu 1998, p. 81).

3.9 'Cleverness' and Victor's Science/Physics Identity

As Bourdieu reminds us, the cultivation of the habitus is a long process. Dispositions are produced through the interplay of different forces and the dialectic between habitus, capital and field:

> continuously defined and redefined in the dialectic between the objectifying intention and the already objectified intention [...], constituted through the confrontation between questions that only exist in and for a mind armed with a particular type of schemes and the solutions obtained through the application of these schemes (Bourdieu 1990, p. 55).

Our analysis suggests that Victor's successful consolidation of a physics habitus and trajectory also required him to navigate/negotiate 'cleverness' – which, as we have written previously, is an inherently classed, gendered and racialized construct that is dominantly aligned with middle-classness, whiteness and masculinity (Archer and Francis 2007). As a white, middle-class boy, Victor enjoyed a privileged structural position that made his identification with cleverness easier than for Other students. However, as discussed next, he still experienced difficulties and challenges in maintaining a viable and legitimate alignment with cleverness, and this relationship shifted and changed over time, requiring considerable identity work and deployment of capital in order to maintain Victor's possibility as 'clever'.

As we have written previously, science in general, but physics in particular, is strongly aligned with notions of cleverness (Archer et al. 2012a, b, 2013, 2020). Indeed, our research shows that physics is widely recognised by students as being the 'most difficult' science. In England, school physics also exercises forms of gatekeeping (such as grading examinations more severely and setting higher entry criteria in comparison to other subjects) which both reinforce and produce the alignment with cleverness. Hence, we argue that for Victor to develop a physics habitus and trajectory, he had to negotiate a viable relationship with 'cleverness'.

At each interview from Year 6 to Year 11, we asked Victor the same question ("Do you think you have to be clever to be into science?") As can be seen below, his responses changed over time:

> Y6: "You don't have to be clever to do science"
> Y8: "I think you have to be a little clever because you have to know about Science in the first place, you have to want to study the subject, yeah, you probably have to be quite clever in the subject to want to learn about it"
> Y9: "People keen on Science … um they're sort of … they're not average people, they're more … they're more clever, they're cleverer than most people"
> Y11: "Er, yeah, you need it, yes. Mainly because sort of you can't just enter Science without knowing what's really happening I think. If you're going to go into Science you need to know what's going on and to know what's going on you need to learn and sort of and then in order to learn you need to sort of either sort of revise or just technically, naturally be clever, sort of".

We interpret the above examples as showing a pattern in which, from an early more egalitarian view (in which "you don't have to be clever to do science"), Victor increasingly aligns science notions of cleverness (whereby one needs to be "naturally clever"). As such, we might view the above pattern of changing responses as hinting at the cultivation of his habitus, through interaction with the field of physics. The importance of cultivating such dispositions is, according to Bourdieu, related to the reproduction of elite institutions, of which, we would argue, physics is a prime example:

> An institution […] is complete and fully viable only if it is durably objectified not only in things, that is, in the logic transcending individual agents, of a particular field, but also in bodies, in durable dispositions to recognize and comply with the demands immanent in the field (Bourdieu 1990, p.58).

Victor's changing views are not uncommon and similar patterns were found among many of his peers in the wider data set. We interpret these shifts as reflecting pedagogic action by the field, in which physics is strongly aligned with notions of cleverness, as part of the reproduction of the subject's elite status. While the link between science and 'cleverness' reflects a wider dominant societal stereotype of science – which we see as playing an important role in the reproduction of science's elite status – as his Year 9 interview highlights, it also required Victor (in his development of a science/physics habitus) to personally negotiate a relationship with 'cleverness' in order to be recognised (and to self-recognise) as a legitimate potential scientist.

Early on in his interviews, despite his view that "you don't have to be clever to do science", Victor identifies himself and is identified by others (notably his mother) as being 'clever'. As Sam put it in the Year 6 interview, "I think he's a very clever kid". However, as Victor starts to develop the disposition that a person needs to be clever to do physics, he also experiences challenges to his own identification as such, which require navigation. These negotiations were exemplified his Year 9 interview, when Victor struggled to reconcile his science habitus with his location in the second set for science:

> Enjoying it [science] has gone down a bit because like there's … I've always been in second set Science, I don't know why I haven't been in top set but it's like I was thinking oh yeah,

> I got a 7 in ... 7A, 8C ... I think I got an 8C in one of the Sciences and I was like 'Oh yeah, that should get me up to the top set' and then I didn't ... I don't know why I didn't go up to the next set. And so I was ... I didn't know what was going on for a long time and then apparently my mum had spoken to ... I think it was [name of science teacher] ... but he said the second set should be the top set but then there's special people in the year that can really excel. But I was just thinking but I'm really doing well in Science as well so I sort of got the feeling that people in the top set were super-geniuses and the people in the second set were just clever.

As the above extract suggests, Victor felt profoundly confused around this time with his location in the second set for science – which sat at odds with his sense of self (as a 'clever' student) and other alleged markers of his cleverness (such as his attainment). Sam also became involved in these negotiations (going into school to speak to the science teacher) and, in the end, Victor negotiated an understanding that enabled him to align himself with cleverness by positioning the top set as exceptions ('super-geniuses').

In his Year 11 interview, Victor reproduced a dominant physics discourse of the 'effortless/naturally clever physicist', that we found to be common among A level physics students across the wider sample ("If you're going to go into Science ... you need to sort of either sort of revise or just technically, naturally, be clever"). As noted earlier, Sam's strong discourse of 'natural' science talent fits well with his alignment with 'natural cleverness':

> I don't know sort of how natural cleverness comes round ... but I've always been told that I'm naturally clever in sort of logical understanding of things. But I do think that you need to be, have some base knowledge or pretty much base knowledge of Science itself, but then for the particular aspects of Science you need specialist qualities (Year 11).

In Year 13, Victor had to negotiate another major challenge concerning 'cleverness' which threatened his continued physics trajectory, when his school did not predict him to attain the top (A/A* grade) A level results. At this point, Sam played a key role in developing and enacting a strategy to navigate the challenge, which involved systematically sifting through every HEIs physics degree entry requirements (alongside exploring non-degree astrophysics apprenticeship routes). As Sam explained:

> So we've looked at universities that do Astrophysics, we've tried to eliminate any universities that want A*s ... we wrote down every single university that did Astrophysics. We ruled out anything that wanted A*s and As (Sam, Year 13)

This strategy paid off and Victor gained access to a university astrophysics degree. Alongside the practical work that Sam put in to securing his continued trajectory, Victor also engaged in focused identity work, to possibilise himself (as a non-A grade student) as a potential physics degree student. For instance, he recounted a discussion that he had held with a physics lecturer whom he had met on a university open day, where lower grades were re-framed in a more positive light:

> I was speaking to one of the people, one of the heads in the Physics department and they were talking about how all the A*, A, A people, quite a lot of them tend to drop out instead of the people that got B, B, C. [...] I found that quite interesting (Victor, Year 13).

Bourdieu described habitus as the 'art of inventing'. We suggest that Victor's data hint at how this art of invention plays out in practice, through strategic manoeuvres (to support and facilitate the required attainment and access to physics courses and routes) and through the identity work – but specifically, the negotiations of a 'clever' identity – that are required to remain a possible/viable scientist/physicist.

3.10 Discussion and Conclusions

> Nothing is more misleading than the illusion created by hindsight in which all the traces of a life [...] appear as the realization of an essence that seems to pre-exist them (Bourdieu 1990, p.55)

In this chapter we have argued that Victor's physics identity and trajectory is the not the sole result of 'chance' or 'mistake' – nor is it the inevitable product of an essentialised or innate talent or aptitude. Rather, our Bourdieusian inspired analysis suggests that Victor's trajectory was the product of lots of hard (conscious and unconscious) work by him and his family, which included the strategic deployment of capital and revealed how the field of physics demands, and achieves, the cultivation of a particular 'clever', 'geeky' masculine habitus/hexis. In particular, we suggested that not only is physics associated with masculinity (and specifically 'geeky' and 'genius' masculinity), but that Victor's successful physics trajectory is realised through the deployment of both capital and the cultivation of the 'right kind' of embodied physics masculine habitus – and that these, in turn, structure Victor's perception of what is possible and desirable for boys like him. The implication of our analysis is that the dominant alignment of physics with a particular, socially privileged, narrow form of masculinity renders it not only a challenge for Victor to possibilise himself, but also, by implication, substantially restricts the possibility of Others (especially those who perform femininity) to be legitimate physics students.

Our application of this analytic lens suggests that it was, in many ways, unsurprising that Victor went on to pursue a STEM degree, given the volume and nature of his science-related family habitus and capital. We interpret the data as revealing a consistent trajectory over the years, involving the cultivation of a distinctive physics habitus – but this process requires concerted work, resourcing, strategizing and navigation by Victor and his family. The mere possession of capital was insufficient on its own. Victor's trajectory also required realisation through the negotiation, cultivation and external recognition of a specific masculine, 'clever' habitus.

To return to Victor's analogy with which we opened the chapter, Victor's astrophysics identity and trajectory required many 'fuses' and repeated 'sparks'. We interpret the ongoing interaction of fuses and sparks as produced through interactions of habitus and capital with the fields of science, physics and the subfield of school physics. As Victor himself suggested at the age of 10, in many ways, Victor had always 'known' that he would study science at university – although his trajectory was not always certain and required considerable effort, work and deployment of capital in order to realise this ambition. Moreover, we suggest that his eventual success (being accepted on to an astrophysics degree course) was strongly facilitated by the web of multiple fuses and sparks which surrounded him. We hypothesise that his successful trajectory would have been much more uncertain had he only had a single fuse and/or spark. His success is also underscored by the cultivation of a habitus that is normalised and demanded by physics. That is, his 'different', academic, 'geeky' masculinity and his achieved alignment with cleverness (though the latter was often threatened and required negotiation), was cultivated through his identification with, investment in, and desire for the field (what Bourdieu would term, *illusio*). Victor's viable physics habitus was affirmed/valued and reinforced through his encounters with physics, and, in turn, we argue, now plays a part in the ongoing constitution and reproduction of the elite nature of the field. In other words, Victor's trajectory was both enabled and possibilised by specific configurations of habitus and capital and by his conformity to the dominant notion of the masculine, 'clever', 'geeky' physicist. Now, as a physicist himself, Victor's embodiment of the demands of the field contributes, in turn, to the ongoing reproduction of the 'reality' and legitimacy of this association.

While Bourdieu argues that economic capital is 'at the root of all the other types of capital' (Bourdieu 1986, p. 24), from our analysis we suggest that Bourdieu's prioritisation of social class obscures the simultaneous importance of gender and 'race' in the social reproduction of physics. While we have not had the space to consider race/ethnicity here, we suggest that our analyses point to an underlying pedagogic action which is aimed at the reproduction of gender privilege (specifically masculinity) in and through the reproduction of physics.

We suggest that the application and integration of a feminist Bourdieusian lens provides additional insights to those provided by a Butlerian lens alone. First, it highlights the importance of capital for enabling, possibilising and realising performances of physics identity and a viable physics trajectory. Second, Bourdieu's concept of field (in interaction with habitus and capital) reveals how particular practices are key for determining the scope and viability of performances of physics identity, particularly, highlighting the slow but inexorable, long-standing cultivation of a physics habitus – which extends powerfully beyond momentary identity performances. In other words, we suggest that the combination of lenses helps us to keep hold of both structured agentic nature of Victor's gendered physics identity and trajectory.

To conclude, we suggest that Victor's case helps us to understand why existing patterns of participation in post-compulsory physics remain low, patterned and resistant to change. That is, inequalities in physics participation are not simply due to differential levels of interest in physics. Rather, they can be understood as produced by the field of physics itself, which requires particular (and demanding) configurations and negotiations of habitus, capital and masculinity in order to maintain a physics trajectory. It could be said that there are only so many Victors in the world who can embody and possess these requirements. We interpret our analyses as also suggesting that if we are to disrupt and change these patterns, then the key will be to change the field – notably the dominant norms, values and habitus that it demands for its elite reproduction. This will mean disrupting dominant associations of physics with masculinity and cleverness. Such an endeavour is, of course, far from simple, and would need to extend across multiple fields, from the popular media (and representations of physics and physicists) through to educational practices (e.g. addressing the current grade severity in Advanced level physics in England) and the culture of school and post-compulsory and professional physics. With regard to the latter, we suggest that one small, but potentially significant, marker of change might be that teachers and students describe and view physics as no 'harder' than any other subject.

References

Adkins, L. (2004). Introduction: Feminism, Bourdieu and after. *The Sociological Review, 52*, 1–18. https://doi.org/10.1111/j.1467-954X.2005.00521.x.

Archer, L., & Francis, B. (2007). *Understanding minority ethnic achievement*. London: Routledge.

Archer, L., DeWitt, J., Osborne, J., Dillon, J., Willis, B., & Wong, B. (2012a). "Balancing acts": Elementary school girls' negotiations of femininity, achievement, and science. *Science Education, 96*(6), 967–989. https://doi.org/10.1002/sce.21031.

Archer, L., DeWitt, J., Osborne, J., Dillon, J., Willis, B., & Wong, B. (2012b). Science aspirations, capital and family habitus: How families shape children's engagement and identification with science. *American Educational Research Journal, 49*(5), 881–908. https://doi.org/10.3102/0002831211433290.

Archer, L., DeWitt, J., Osborne, J., Dillon, J., Willis, B., & Wong, B. (2013). Not girly, not sexy, not glamorous: Primary school girls' and parents' constructions of science aspirations. *Pedagogy, Culture & Society, 21*(1), 171–194. https://doi.org/10.1080/14681366.2012.748676.

Archer, L., Dawson, E., DeWitt, J., Seakins, A., & Wong, B. (2015). Science capital: A conceptual, methodological, and empirical argument for extending Bourdieusian notions of capital beyond the arts. *Journal of Research in Science Teaching, 52*(7), 922–948.

Archer, L., Moote, J., Francis, B., DeWitt, J., & Yeomans, L. (2017). The "exceptional" physics girl: A sociological analysis of multimethod data from young women aged 10–16 to explore gendered patterns of post-16 participation. *American Educational Research Journal, 54*, 1–39. https://doi.org/10.3102/0002831216678379.

Archer, L., Moote, J., & MacLeod, E. (2020). Learning that physics is 'not for me': Pedagogic work and the cultivation of habitus among advanced level physics students. *Journal of the Learning Sciences*, 1–38. https://doi.org/10.1080/10508406.2019.1707679.

Bourdieu, P. (1986). The forms of capital. In J. G. Richardson (Ed.), *Handbook of theory and research for the sociology of education* (pp. 241–258). New York: Greenwood Press.

Bourdieu, P. (1990). *The logic of practice*. Cambridge: Polity Press.

Bourdieu, P. (1993). *The field of cultural production: Essays on art and literature*. Cambridge: Polity Press.

Bourdieu, P. (1998). *Practical reason*. Cambridge: Polity Press.

Bourdieu, P. (2001). *Masculine domination*, R. Nice (Trans). Cambridge: Polity Press. Originally published as La Domination Masculine. Paris: Seuil.

Bourdieu, P. (2010). *Sociology is a martial art*. New York: New Press.

Bourdieu, P., & Wacquant, L. J. D. (1992). *An invitation to reflexive sociology*. Cambridge: Polity Press.

Butler, J. (1990). *Gender trouble*. Routledge, Chapman & Hall Inc: London & New York.

Carlone, H. B. (2003). (Re)producing good science students: Girls' participation in high school physics. *Journal of Women and Minorities in Science and Engineering, 9*, 17–34. https://doi.org/10.1615/JWomenMinorScienEng.v9.i1.20.

Carlone, H. B., & Johnson, A. (2007). Understanding the science experiences of successful women of color: Science identity as an analytic lens. *Journal of Research in Science Teaching, 44*(8), 1187–1218. https://doi.org/10.1002/tea.20237.

Danielsson, A. T. (2012). Exploring woman university physics students "doing gender" and "doing physics". *Gender and Education, 24*(1), 25–39. https://doi.org/10.1080/09540253.2011.565040.

DeWitt, J., Archer, L., & Mau, A. (2016). Dimensions of science capital: Exploring its potential for understanding students' science participation. *International Journal of Science Education, 38*(16), 2431–2449. https://doi.org/10.1080/09500693.2016.1248520.

Dika, S. L., & Singh, K. (2002). Applications of social capital in educational literature: A critical synthesis. *Review of Educational Research, 72*(1), 31–60. https://doi.org/10.3102/00346543072001031.

Francis, B. (1999). Lads, Lasses and (New) Labour: 14-16-year-old students' responses to the 'laddish behaviour and boys' underachievement' debate. *British Journal of Sociology of Education, 20*(3), 355–371. https://doi.org/10.1080/01425699995317.

Francis, B. (2000). The gendered subject: Students' subject preferences and discussions of gender and subject ability. *Oxford Review of Education, 26*, 35–48. https://doi.org/10.1080/030549800103845.

Gonsalves, A. (2014). "Physics and the girly girl—There is a contradiction somewhere": Doctoral students' positioning around discourses of gender and competence in physics. *Cultural Studies in Science Education, 9*, 503–521. https://doi.org/10.1007/s11422-012-9447-6.

Israel, G. D., Beaulieu, L. J., & Hartless, G. (2001). The influence of family and community social capital on educational achievement. *Rural Sociology, 66*, 43–68. https://doi.org/10.1111/j.1549-0831.2001.tb00054.x.

Lareau, A. (2003). *Unequal childhoods: Class, race and family life*. Berkeley: University of California Press.

Lemke, J. L. (1990). *Talking science: Language, learning, and values*. Westport: Ablex.

Martin, N. D. (2009). Social capital, academic achievement, and postgraduation plans at an elite, private university. *Sociological Perspectives, 52*, 185–210. https://doi.org/10.1525/sop.2009.52.2.185.

McLeod, J. (2005). Feminists re-reading Bourdieu: Old debates and new questions about gender habitus and gender change. *Theory and Research in Education, 3*(1), 11–30. https://doi.org/10.1177/1477878505049832.

McNay, L. (1999). Gender, habitus and the field: Pierre Bourdieu and the limits of reflexivity. *Theory, Culture & Society, 16*(1), 95–117. https://doi.org/10.1177/026327699016001007.

Mendick, H., & Francis, B. (2012). Boffin and geek identities: Abject or privileged? *Gender and Education, 24*(1), 15–24. https://doi.org/10.1080/09540253.2011.564575.

Mendick, H. (2016). Gender and physics: A sociological approach. *Physics Education, 51*(5), 1–6. https://doi.org/10.1088/0031-9120/51/5/055014.

Moi, T. (1991). Appropriating Bourdieu: Feminist theory and Pierre Bourdieu's sociology of Culture. *New Literary History, 22*(4), 1017–1049.

Moote, J., Archer, L., DeWitt, J., & MacLeod, E. (2019). Comparing students' engineering and science aspirations from age 10 to 16: Investigating the role of gender, ethnicity, cultural capital, and attitudinal factors. *Journal of Engineering Education*, 109. https://doi.org/10.1002/jee.20302.

Ong, M. (2005). Body projects of young women of color in physics: Intersections of gender, race, and science. *Social Problems, 52*, 593–617. https://doi.org/10.1525/sp.2005.52.4.593.

Osborne, J., & Dillon, J. (2008). *Science education in Europe: Critical reflections. A report to the Nuffield Foundation*. London: King's College London.

Perna, L. W., & Titus, M. A. (2005). The relationship between parental involvement as social capital and college enrolment: An examination of racial/ethnic group differences. *The Journal of Higher Education, 76*(5), 485–518. https://doi.org/10.1080/00221546.2005.11772296.

Sandefur, G. D., Meier, A., & Campbell, M. (2006). Family resources, social capital and college attendance. *Social Science Research, 35*, 525–553.

Vincent, C., & Ball, S. (2007). *Childcare, choice and class practices*. London: Routledge.

Webb, J., Schirato, T., & Danaher, G. (2002). *Understanding Bourdieu*. Sydney: Allen & Unwin.

Chapter 4
An Intersectional Physics Identity Framework for Studying Physics Settings

Angela Johnson

4.1 Introduction

Women of color are markedly underrepresented among people who receive bachelor's degrees in physics in the United States. Less than 4% of physics bachelor's degrees awarded between 2004 and 2014 went to Black, Latina, Asian, American Indian and mixed-race women, although they make up 16% of the people who received bachelor's degrees in any subject during that time, and 22% of the US population age 18–24 (National Science Foundation and National Center for Science and Engineering Statistics 2017). Furthermore, there is reason to think that women of color who come to college planning to major in physics are more likely to change majors than male students or White students; women completed STEM majors in 2013 at lower rates than they declared interest in STEM in 2007; this was also true for students from all non-White racial groups (National Science Foundation 2016; see figures 2–13 and 2–14).

Qualitative research suggests that the few women of color who pursue physics majors experience discouraging, alienating conditions. In a study of physics majors that included 10 Black women and Latinas, Ong (2005) found that "regardless of their actual abilities as measured by exam performances, grade point averages, and research mentor evaluations, women of color participating in the study said they perceived nearly consistent messages – with some rare exceptions – that because they lack the standard appearance of a scientist, they also lack the intellectual competence associated with such an appearance" (p. 602). One African American woman told Ong "I've noticed [my peers] in their tone of voice that they take with me … They feel the need to explain things that much more because, well, this Black person won't get it. I see them doing it with Latino students and doing it to Black

A. Johnson (✉)
St. Mary's College of Maryland, St. Mary's City, MD, USA
e-mail: acjohnson@smcm.edu

A. J. Gonsalves, A. T. Danielsson (eds.), *Physics Education and Gender*, Cultural Studies of Science Education 19,
https://doi.org/10.1007/978-3-030-41933-2_4

students. They go into this extra detail: 'Do you understand? Do you understand?' – as though my intellect is gonna be different from someone who's Asian or White" (p. 602). In another related paper, Ko and colleagues quote Elena, a Latina physicist: "In physics, as a woman of color, you will not find the way paved... You walk into a stranger's territory, people look at you, you look and perceive yourself to be different, you are walking the narrow path" (Ko et al. 2014, p. 186).

In a longitudinal study involving 17 women physics majors and physicists of color, 13 talked about feeling isolated in physics settings, and 15 reported experiencing microaggressions (defined as "subtle indignities, slights, or insults directed at individuals, consciously or unconsciously, because of their race or gender"; Johnson et al. 2017). Other research has demonstrated the perniciousness of microaggressions. In a study of 21 predominantly White women who were pursuing PhDs in physics and astronomy, 16 reported instances that the researchers categorized as microaggressions, and 5 reported instances of hostile sexism (Barthelemy et al. 2016). The researchers described microaggressions as "subtle forms of discrimination that are often socially engrained and unconscious. An example of a gender microaggression would be not listening to a woman's idea but then responding to the same idea from a man, or not thinking to initiate a collaborative project with a woman" (p. 4). Tolerating microaggressions and hostile sexism resulted in "a physical and cultural environment that is discouraging of women's participation in physics and astronomy" and "ignoring these women's ideas, conveying a message of women as objects, and restricting access to laboratory equipment" (p. 11).

However, I have been carrying out research in a setting where the three women physics majors of color had generally positive things to say about their experiences in physics. A Black third year student, when I asked her how she thought about the dearth of Black women in physics in general, said "physics is what I've always been interested in. It doesn't feel like I'm out of place. It's the subject I'm interested in. So I don't really think about it." When I asked a mixed race fourth year student what it was like to major in physics at this institution, she told me "I mostly love it. It's been a really great experience majoring in physics here. It's been really really hard but I love it!" Of course given the small sample size in this study, it's possible that their experiences were anomalous; however, I spent considerable time immersed in this department, and I think they are making such positive statements because the culture is indeed less toxic for women of color than is the case in most physics departments. White women in the department made similar statements, describing their academic faculty as "friendly" and "helpful" and contrasting their experiences with less-supportive experiences they had had in other physics settings.

4.2 Intersectional Physics Identity

I have written elsewhere, with wonderful co-authors, about the identity of women of color majoring in science at predominantly White institutions (Carlone and Johnson 2007; Johnson et al. 2011). In this study, I conceptualize identity not as an internal individual experience but as *a feature of a social setting*. I consider questions like: What do personal interactions, cultural features and structures in the setting convey about what kinds of people belong here? To clarify: I think a person may have an affinity with physics—may self-identify as a "physics person," someone who enjoys physics and could be good at it. But an individual's affinity with physics doesn't necessarily shed light on the kinds of identity that are supported in a particular physics setting. Affinity to physics might make a person more likely to want to major in physics (although I can imagine cases where a person would major in physics without having an affinity for it) but it's not what I'm interested in here. I'm interested in what happens once people have chosen, with or without an affinity for physics, to try to join a physics setting.

This framing of identity works well with an intersectional analysis. The central idea behind intersectionality theory is that different kinds of people may experience the same setting differently depending on various personal characteristics. Women may experience a physics setting differently from men, but it's not as simple as that; Black women may experience it differently from White women (or from Black men) or women of any other race or ethnicity; affluent women may experience it differently than women who struggle financially; religion may affect experience, too, and sexuality, and any other dimension along which humans can be divided. This is where the term comes from – the idea being that we all exist at particular intersections within social settings, and the intersections we exist at affect how we are perceived and our access to power in those settings (Crenshaw 1989, 1991). The salience of the dimensions that define our particular intersection change depending on the setting; for a Black woman, majoring in physics at a predominantly White university means something different from majoring in ethnic studies at the same university or majoring in physics at Spelman (a historically Black women's college[1]).

But it's not just that one's experiences differ according to setting and social location; these "distinctive social experiences" that result from intersecting systems of power "are fundamentally unjust" (Collins 2015, p. 14). People at different intersections have access to different opportunities; their words and insights are given different levels of attention and credibility; they are welcomed, overlooked or actively dismissed. "At the same time that structures of race, class, and gender create disadvantages for women of color, they provide unacknowledged benefits for those who are at the top of these hierarchies" (Zinn and Dill 1996, p. 327). It's safe to say that in most physics settings, power relations are such that the status quo is maintained; White men find it easier to thrive, have access to more opportunities, receive more attention, find more social support. They find it easier to inhabit an available physics identity. To look at this issue from the perspective of women of color, "the performance of different identities creates higher levels of stress for individuals who belong to more than one group that is underrepresented in physics" (Traxler et al. 2016, p. 9).

[1] Historically Black colleges and universities (HBCUs) are institutions of higher education in the U.S. that were founded before the U.S. Civil Rights Act of 1964. Before this legislation, many U.S. higher education institutions were not open to Black students; HBCUs were founded to provide Black students with access to university educations. Although HBCUs are now technically integrated, they still have the education of Black students as their primary mission.

My interest is in understanding a physics department where women of color feel successful, feel like they belong; in identity terms, I want to understand the physics identity which is available in this setting, and how that identity is accessible to women of color. This commitment is congruent with the origins of intersectionality, which are political, not scholarly. The ideas that we now call intersectionality were developed by women of color and other activists over decades, especially in the 1960s–1980s (see, for example, Anzaldúa 1999; Guy-Sheftall 1995; Moraga and Anzaldúa 1981; Smith 1983). Collins makes it clear that the goal of this work was not a search for "the latest theoretical innovation" (Collins 2015, p. 8). Rather, "Black feminism's immediate concern in the United States was to empower African American women through critical analyses of how mutually constructing systems of oppression of race, class, gender, and sexuality framed the social issues and social inequalities that Black women faced.... Their intersectional framework suggested provocative links that might ground social justice projects" (p. 8–9). It is intersectionality's power as a tool to promote social justice that makes it a valuable framework for this study; intersectional analyses can help us both identify the features of physics settings that make them more challenging for women of color to navigate than other kinds of people (for a particularly strong example of this, see Kachchaf et al. 2015) and, as in this study, identify features of a setting that make it more amenable to women of color. Using intersectionality with the intention to advance social justice is what Collins calls "critical praxis:" "knowledge projects that take a stand; such projects would critique social injustices that characterize complex social inequalities, imagine alternatives, and/or propose viable action strategies for change" (Collins 2015, p. 17). It is my goal in this particular project to help us imagine alternatives: How physics settings could be structured so that women, particularly women of color, would belong more easily in those settings. What would a physics setting be like where valued identities in that setting could be assumed as easily by women of color as by White men?

Collins (Collins 2009; Collins and Bilge 2016) provides another analytic tool in addition to intersectionality to help with this quest. She argues that if we are interested in understanding how power is organized in a setting (which is to say, who belongs in the setting, who has access to the opportunities in the setting), we need to consider four domains: the interpersonal, the cultural, the disciplinary and the structural; she calls this the Domains-of-Power Framework (Collins 2009).

In the *interpersonal domain*, power is expressed between individuals; a person may be included or excluded, praised or criticized, honored or overlooked. It is the "domain of one-on-one encounters and the area of personal choice" (p. 53–54). In my prior research on the experiences of women of color majoring in STEM, one pattern of interaction which made it more difficult for women of color to be recognized as good science students was the choosing of lab partners. A number of women told me about how White students would immediately partner up and leave them either partnerless or working with the few other students of color in class (Johnson 2007). An example of how personal choice can determine who is valued in a setting comes from a study in which male and female biology, chemistry and physics faculty, at all career stages, were asked to evaluate the application materials of a hypothetical applicant for a lab job (Moss-Racusin et al. 2012). Each participant received the same application materials; the only difference is that some had a male name and some had a female name. Participants rated the male applicant significantly higher and indicated that he should be paid more; this was true after controlling for age, gender and discipline – so women scientists were doing it too. In the study I am presenting in this chapter, given the small, bounded setting (one physics department, consisting of a handful of academic faculty and 30 declared majors), I focused on how students interacted with one another, how faculty interacted with students, and how faculty interacted with one another. I gathered data both from my own direct observations and from what members of the setting told me.

The *cultural domain* is where a group's values are conveyed (or contested); "when it comes to the organization of power, ideas matter in providing explanations for social inequality and fair play" (Collins and Bilge 2016, p. 10). The cultural domain is very powerful in physics settings; US culture is saturated with what it means to be a physicist. Physicists are eccentric geniuses, experiencing flashes of insight, laboring in solitude. Researchers analyzing case studies of laboratory practices in three physics settings found that "a static and nonperformative idea of masculinity may be associated with many of the valued attributes also associated [with] being a physicist" (Gonsalves et al. 2016, p. 4). In an analysis of interviews from 21 women pursuing graduate work in physics and astronomy, researchers documented "a physical and cultural environment that is discouraging of women's participation in physics and astronomy….[which] resulted in ignoring these women's ideas, conveying a message of women as objects, and restricting access to laboratory equipment" (Barthelemy et al. 2016, p. 11). Put more simply, physics students, as a friend of mine told me decades ago, "don't shower much." They are Sheldon Cooper, solitary, living his life according to bizarre systems; Newton, developing the theory of universal gravitation in one glorious moment after being hit on the head by an apple; Einstein of the crazy hair and endearing accent. Of course the cultural domain also influenced the outcomes of the lab applicant study; academic faculty were exercising personal choice when evaluating the fictitious application, but that choice was influenced by cultural ideas about who is good at science.

The *structural domain* has to do with how power is allocated via large social structures. When writing about colorblind racism, Collins talks about "how racial practices are organized through social institutions such as banks, insurance companies, police departments, the real estate industry, schools, stores, restaurants, hospitals and governmental agencies" (Collins 2009, p. 53). An example of how the structural domain plays out in STEM is the pipeline issue in the US: many poor students and students of color can never even enter the pipeline, because they attend schools where there is no access to high-level math and science classes. "A quarter of [US] high schools with the highest percentage of black and Latino students do not offer Algebra II; a third of these schools do not offer chemistry" (Office for Civil Rights 2014, p. 1). In this study I am looking at structures on a smaller scale; the structure of the major itself – how physics classes are officially organized. When analyzing the 2014 FIFA World Cup, Collins and Bilge used the structural domain in this way; they specified that the structural domain in that study referred to "how FIFA itself is organized or structured" (Collins and Bilge 2016, pp. 11–12).

The *disciplining domain* focuses on how rules and regulations get enforced, and for whom; whether, as Collins said when writing about colorblind racism, "people use the rules and regulations of everyday life to uphold the racial hierarchy or to challenge it" (Collins 2009, p. 53). Although Collins called this the disciplinary domain, for the purposes of this study I am calling it the disciplining domain, to avoid the confusion between disciplinary as referring to the discipline of physics and disciplinary as referring to the disciplining (regulating, punishing) of members of a physics setting. An example of how power gets allocated in the disciplining domain of STEM comes from the finding that when grant applications were not blinded, Black applicants were 10% less likely to receive NIH grants than White applicants, after controlling for factors like educational background, previous research awards, and publication record (Ginther et al. 2011). The rules for awarding grants were apparently used differently when evaluating Black scientists. Again in this example we can see the interpersonal and cultural domains at work as well; the way grant reviewers used the rules for awarding grants was influenced by their cultural understanding about who is good in science and resulted in them making personal choices about how to score each grant. The power of the disciplining domain is illustrated by the experiences of the five women who reported hostile sexism to Barthelemy et al. (2016). "In the cases of all of these women, four attempted to find resolution by reporting their experiences to superiors. However, with only one exception (Melissa), these reports were met with deaf ears" (p. 11).

At this point it is useful to pause and think about the prototypical physics department; the physics department that exists in our shared imagination. In this department, women are isolated; women of color even more so (National Science Foundation and National Center for Science and Engineering Statistics 2017). They may experience microaggressions on top of that isolation (Barthelemy et al. 2016; Johnson et al. 2017). The department culture is marked by competition and the belief that physics is a meritocracy, students with the most natural ability will deservedly succeed, and difficulties are an indication that a student isn't cut out for physics. Physics classes consist of faculty lecturing up at the front of the room; high levels of student failure are expected. In the disciplinary domain, it's the Wild West, the boy's club; faculty don't intervene in how students interact with one another (Table 4.1).

It's hard to know just how common this prototypical physics department is, but we all immediately recognize it (it's the reason that I attended a women's college decades ago). It's easy to understand why there would be so few women of color (or women at all) in a prototypical physics department, because power relations are structured at each level so that women face obstacles to being recognized as belonging in the setting – obstacles that White men don't face. The interpersonal domain would be so discouraging; who wants to endure years of isolation and subtle digs at one's dignity and personhood? The emphases in the cultural domain on competition and natural genius work against women, too. As Seymour and Hewitt pointed out decades ago (Seymour and Hewitt 1997), women and men interpret the competitive, alienating environment of most STEM classes differently. Whereas men are socialized to be self-sufficient and stoic, women are socialized to "perform for the

Table 4.1 The Domains-of-Power Framework in prototypical physics departments

	Typical physics departments
Interpersonal domain	Women of color experience isolation and experience microaggressions.
Cultural domain	Physics is a competition; good physicists are natural geniuses who work in isolation.
Structural domain	Classes consist of faculty lecturing; high levels of student misconception and failure are expected.
Disciplining domain	Faculty do not intervene in student-student interactions

approval of others" (p. 266). Thus, "in treating male and female students alike [by subjecting both to the competitive environment in STEM], faculty are, in effect, treating women in ways that are understood by the men, but not by the women" (p. 262). The structural domain, with its emphasis on delivery of information rather than on student learning, does nothing to offset the unpleasant conditions in physics; nor does the hands-off approach that prototypical physics faculty take to conditions in physics settings. This prototypical physics department serves as a useful foil for analyzing the components of physics identity in real departments.

4.3 Methods

I used Collins' Domains-of-Power Framework to develop both a set of guidelines for gathering data about physics identity, and also a set of questions to help me analyze that data intersectionally – to consider how that identity aligns with and how it contests wider patterns in physics and in society; see Table 4.2.

These questions allowed me to develop a sketch of who belongs in the physics department I was studying – the ways members of the department interact with one another, the cultural beliefs of the setting, the values and goals embedded in the setting's structures, and what kinds of actions and speech, on all these levels, were considered important enough that people who violated them were reprimanded. This theoretical framing lent itself beautifully to the traditional interests of anthropology: What people say, what they do, and the material objects they use, and most especially the meanings they share about these. The traditional ethnographic methods – participant observation and open-ended interviews – can be used to gather the kind of nuanced information this framework demands (Spradley 1979, 1980).

Thus, I immersed myself in the life of the physics department under study. I attended at least one class session of every physics class and lab being taught during the semester I began the study, with the exception of the one class that was being offered only for non-physics majors. I made sure to attend the 100-level class as well as an Emerging Scholars-type seminar (Fullilove and Treisman 1990; Treisman 1992) several times during the first 2 weeks of the semester, so that I could observe what happened as new students attempted to negotiate the setting for the first time.

Table 4.2 Questions to support an intersectional analysis of physics identity

Domain	To look for	To consider
Interpersonal	How do students interact with one another?	Are there patterns along race and gender lines? Do they conform to or contest common patterns in physics? In society?
	How do students interact with faculty?	
	How do faculty interact with students?	
	How do faculty interact with one another?	
Cultural	What do the words and actions of faculty convey about what's valued in the domain?	How do the things that are valued align with larger cultural beliefs about race and gender? Conform to or contest common patterns within physics settings? In society?
	What do students' words and actions convey about what's valued in the setting?	
	What does the setting itself – the objects which make it up – convey about what's valued in the setting?	
Structural	What are the policies in classrooms, labs and other physics spaces?	Do the policies in place serve to perpetuate or challenge under-representation in physics?
Disciplining	What kinds of student behaviors do faculty correct?	How do the student actions that faculty condemn or correct align with or challenge under-representation in physics?
	What kinds of student behaviors do students believe faculty would correct?	

When I attended classes, the academic faculty welcomed me and made use of my presence; as students worked on problems during class, I circulated and did my best to help them. At the invitation of faculty, I also spoke about the pleasure and challenge of being a high school physics teacher, and how to become a teacher.

Besides attending classes and labs, I interviewed 6 out of the 8 women majoring in physics, including all three women of color; during interviews, I started by asking them "tell me your life story in physics." For some students, their answers were so wide-ranging that this question was enough; for others, I also asked them "what is your life in physics like?" Typically I also showed them NSF data about the under-representation of women in physics, as well as the somewhat better numbers at their own institution, and asked them what they thought of these patterns. Each interview was highly idiosyncratic as I followed each participant's interests. Interviews typically lasted about an hour. At the end of the interviews, the focus often shifted to the students asking me for guidance about their future choices. At the institution where I carried out this research, faculty advise students intensively, so this was a normal shift in this context. First and second year students tended to talk about pursuing research opportunities; third and fourth years were more focused on graduate school options and careers.

I interviewed four of the five physics faculty members; those interviews typically lasted more than an hour (questions are in Table 4.3, below). In addition to the formal interviews, I spoke repeatedly with the faculty members. I am a faculty member in a teacher education program and a former high school physics teacher; I taught AP physics C (the more rigorous AP physics sequence) for many years and the members of the department I studied often talk with me about educational research design (as a department, they avidly study their own teaching) and about inclusion and diversity issues. For example, one faculty member asked me for feedback about how to encourage more students to participate during class; another one consulted with me about a struggling student.

Finally, I attended a physics class in which most of the third and fourth year students were enrolled, and used it as a focus group. Again see Table 4.3 for the questions I asked. I told the students that the goal was to hear from as many voices as possible, not to reach consensus. For popular responses, I noted how many people agreed. Although the women physics majors in the department where I carried out this study were by-and-large positive about their experiences, 75% of the physics majors were still men, and this focus group led to the curious situation of a group of male students explaining to me what it was like to be a woman majoring in physics, while their female classmates sat quietly and their male faculty member grew increasingly annoyed at them – but the data I gathered from the focus group was nonetheless useful (and the women in class laughed good-naturedly afterwards, for which I commend them).

I coded the data by searching for common patterns, and then using those common patterns to answer the questions in column 2 of Table 4.2. Once I had tentative answers to those questions, I looked through the data both for discrepant examples

Table 4.3 Interview questions

Students during individual interviews	Tell me your life story in physics.
	What is your life in physics like?
Faculty	What does it mean for someone to be a good physics student?
	What do you do as an individual to support and teach students?
	What does your department do?
	Why do you think this institution has these good numbers for retaining women, and what could be done to make them even better?
	For women: What was it like for you to study this field? What has it been like for you to be faculty at this institution?
Students during focus group	Why did you decide to major in physics?
	What was your route to this particular institution?
	Is majoring in physics like you thought it would be?
	What is it like to major in physics here?
	What do you like about physics classes?
	What could be even better?
	Why do you think this institution has more women in physics than average?
	What ideas do you have for attracting even more women?

and for examples where participants contrasted this setting with other physics settings where they had spent time. I carried out the coding using NVIVO. Once I was confident that the answers to my questions were supported by my data, I considered the implications of those answers using the questions in column 3 of Table 4.2.

I shaped this study around Patricia Hill Collins' four precepts of Black feminist epistemology (Collins 2000). Collins argues that "many Black women have had access to [an] epistemology that encompasses standards for assessing truth that are widely accepted among African-American women" (p. 256). These standards are:

- Truth is grounded in experience (so that "those individuals who have lived through the experiences about which they claim to be experts are more believable and credible than those who have merely read or thought about such experiences," p. 257).
- Truth is arrived at through dialogue ("new knowledge claims…are usually developed through dialogues with other members of a community," p. 260).
- Truth demands caring ("personal expressiveness, emotions, and empathy are central to the knowledge validation process," p. 263).
- Truth demands accountability ("Knowledge claims made by individuals respected for their moral and ethical connections to their ideas will carry more weight than those offered by less respected figures," p. 265).

Thus, as the researcher, I saw my role as getting to know people in the setting better and better, rather than seeking objectivity. I am an avowed advocate for women in physics; I want to better understand both the obstacles they face and the characteristics of settings where they thrive, because I want physics to become more inclusive. I see it as my job to behave in ways that my participants can trust and respect, not just in order to gain greater access to the setting but because I would be morally compromised were I to do anything else. In carrying out this study, my interviews with faculty and students were shaped as ongoing conversations, where I checked with them to ensure that I understood what they were saying, and continued (and continue to this day) to talk with them about their thoughts that arise from this project. My research emphasis – and indeed the emphasis of all ethnography – was on experience, and I centered the experiences of the women of color in the setting, always sitting near them while doing participant observation in classes, for example. I expect myself to be accountable to all members of the setting, but first and foremost to the women of color. When I use students' words in public, I first check with them to be sure they are comfortable, particularly the women of color, whose identities it is difficult to obscure. This is on top of the standard consent procedures I made use of; I double-check, and I send out the language I will use in publications, so participants can see for themselves exactly what they are agreeing to. So, for instance, before this chapter went to press, I sent it to all six students I interviewed as well as all four faculty members, indicating to each person where I was using her or his words and asking for (and receiving) clarifications and for permission to use those words. I have continued to advise students about their career choices even past graduation, and I actively put them forward for physics opportunities – for example, connecting a student to the National Society of Black Physicists.

4.4 Physics Identity in This Setting

In all four power domains, the physics identity that I found expressed in this setting was strikingly different from the prototypical physics student identity. Two related forms of physics identity emerged, one for students, the other for academic faculty. To be a good physics student involves doing group work (whether students like it or not), working on physics homework together, and helping other physics students. Good physics students are curious, interested and engaged; they think critically about concepts and abstractions; but they don't have to be on track to pursue a PhD – many career paths are valued. To be a good physics faculty member (in the eyes of a student) involves being accessible to students, seeking shared goals through discussion, using research-based, interactive teaching practices, letting assessment guide teaching practices, emphasizing the importance of hard work (rather than natural ability) in physics; and, for male faculty, understanding the challenges faced by women in physics and addressing those challenges. These components of physics identity were conveyed in all four domains of power, as I discuss.

4.5 Interpersonal Domain

4.5.1 How do students interact with one another?

At the time I carried out this study, there were eight women majoring in physics; I interviewed six of them, including all three of the women of color and mixed race women. Five out of the six women I interviewed spontaneously told me about how the physics majors *work together* and *help one another* – which is especially appreciated because the work is difficult. One woman summed it up like this: "Physics has the past couple years been more difficult, more time-consuming classes. But that was OK because I enjoyed it more anyway. Misery loves company, and there's always company within the physics department!" They used the following words to describe physics majors: "friends," "everybody is so friendly," "super nice!", and "I definitely socialize with all of them." This quote from a second year student pulls all these themes together: "Sometimes I'll be in [the physics building]. So I can just like ask a question. Maybe not my class, but it could be an older physics student that could help me. They're super nice! [Q: just women? or women and men?] Either one. Whoever's there. I think I could ask any of them. We're kind of all in it together, why wouldn't they help me? They know what I'm going through – they've done it themselves!"

Several women told me that when they first started out as physics majors, they found the male-dominated environment intimidating, but this culture of *working together* helped them get over their initial intimidation. A fourth year student told me "now I've made friends with everybody and know everybody, I'm fine with it. But at first it's intimidating…. [Now] it's not just a bunch of men – they're people you know, who you become friends with. So it's not like intimidating anymore." A second year student described it this way: "At first, when I went through the introductory classes, I kind of thought like I didn't have much of a natural ability for physics, but that just wasn't true. As I've gone through it I've kind of noticed how even the guys struggle as much as I do, but I might not have known it before. There's more of a feeling, like, we're sort of all in it together. I don't feel like they know more than me because they're guys or something, or they get it more."

This common practice of working together is a central element of what it means to be a physics student in this setting, and it draws on strengths that are associated with the feminine; it creates a space where women can flourish. And it's not just the women who talk about the physics department this way. These themes came up during the focus group I held with third and fourth year students majoring in physics ("You get to know everyone in the major, and you get to know all the professors;"[2] "You learn what everyone's good at; if you get stuck on something, you know who you can to go ask. Sophomore and junior year are really tough – going through the hardship with everyone – it's nice.") During the focus group, several men talked about how they dislike this aspect of the major (one man said that when he began majoring in physics, "I just didn't really want talk to people. Wanted to sit around in a corner by myself doing cool things"), but recognize its value: "I know it's good for me but I still kind of dislike group work, and the amount of it I have to do, because I'm a pretty antisocial person. I recognize how good it is for my learning, but there are times I just want to copy what's on the board." As soon as he said this, a woman responded "I feel the opposite. I find it more satisfying if I do it with other people."

4.5.2 How do academic faculty interact with students?

Below, in the section where I analyze this setting in the structural domain, I discuss how faculty interact with students formally, in their classrooms. Here I want to report that the students' focus on working together is replicated in how the academic faculty work with the students. Four of the six women physics majors I interviewed spontaneously told me that the *faculty are accessible*, using words like "nice" and "helpful." One told me that her research advisor "is like the nicest professor I've ever met in my life. She's…if you do something wrong, she doesn't even frown at you. She's so helpful, she's just…everyone loves her. Especially – I've heard the general physics kids really like her, and they're the ones who are forced into taking physics and don't actually want to! It seems like she's found a way to really explain things well and get people to be productive without being mean or making you feel bad." Another one told me that *both male and female faculty are accessible*. After we looked together at some statistics indicating that more women

[2] Note that professor is the standard term for any US academic faculty member; thus it is used throughout the quotes in this chapter to denote anyone with an advanced degree teaching a physics class.

study physics at this institution than the national norm, a woman told me "I would love to say it's because we have more female professors than male professors, but that's not true. I feel more comfortable talking to any of those professors. I've never felt intimidated by the male professors more than the female professors." This accessibility is not a coincidence; the faculty make a deliberate choice. One told me "I try and make myself really open to if they have questions – just trying to be around the department, so they can find me and ask me if they have questions." Another said "I try to mentor people, and develop relationships with students and help them to grow and figure out what to do with themselves, and understand that success is what they define it to be and not what someone tells them to be."

Finally, the newest physics faculty member told me that her colleagues offer her the same support: "They've been very good about checking in with me as far as – how are you feeling? how are things going? are there any resources you need from the department? Do we have access to the journals you need for your research? … everyone being completely open to me hopping into their classroom, saying 'here's something I don't know what to do about' [responding with] 'come in, shut the door, I'll tell you my viewpoint and what I think the viewpoint of other people will be, and you can go and ask them.' Being super-generous with their time."

4.5.3 How do academic faculty interact with one another?

When faculty spontaneously told me how they interact with one another, two themes emerged strongly. First, the department *seeks shared goals through discussion*. One told me "I don't want all of us to be thinking the same way about all the problems. What I do want are shared aspirations. I want everyone to have the same goals for the department that I do, but everyone should have their own ideas about what the best ways to achieve them are. I think that's been a great strength of the department. The faculty are pretty markedly different, but I think that all of us share the same aspirations." Another echoed this: "We're going to be as a whole talking about reaching consensus on what students should be learning, and assessing and seeing what changes we need to make in order to better meet the learning outcomes."

An even stronger theme came through what both of the women faculty told me: that their *male colleagues take responsibility for gender issues*. One told me "It's almost as if because someone else is thinking about this, I don't have to waste my brain thinking about it. In other environments, I would ignore that I was the only other woman in the room – no-one else was acknowledging it – I would have to flip that part of my brain on and run it in the background. Here, because someone else has got this, I can just do physics. … We're all handling this together, nobody has to handle this themselves." Notice how she talks about how in other settings she had to run her awareness of her gender "in the background" – a constant drain on her cognitive resources. Her colleague talks about the emotional weight that is lifted because their male colleagues take responsibility for thinking about gender issues:

"[the men] really are very proactive about understanding the issues of being a woman in physics, so I don't have to educate them, which is a huge relief." I saw this in action when one of their male colleagues told me about how another department encourages students to call faculty by their first names. He told me that one of his female colleagues "has raised the point that there are ways that can be somewhat problematic" by playing into the tendency some students already have to disregard the expertise of female faculty. Because of this, he told me, both he and his male colleague refer to the women in their department as "professor" in front of students because of "both the fact that they're women, and also that they're younger."

I would like to end this section with a story about how a student interacted with me. In one class where I was conducting participant observation, the faculty member gave me a few minutes to talk about my project, and when I told the students that the number of women majoring in physics, math and computer science at their college is significantly higher than at other liberal arts colleges, a student cheered. This is a place where you can publicly express your support for women in physics.

4.5.4 Are there patterns along race and gender lines? Do they conform to or contest common patterns in physics? In society?

This is a setting with a high degree of social support and interaction, where the ability to work together, be helpful and be friendly is highly valued for members of the setting, to the point that even people who don't enjoy high levels of interaction see it as something they must engage in to be in the setting. For women of color, especially, this high level of social inclusion is in stark contrast to their typical experience of isolation in physics settings (Johnson et al. 2017). It's a setting where people with more power take responsibility for the success of those with less power (faculty helping students and more junior faculty; more advanced students helping first and second year students). Interactions in the interpersonal domain of power in this setting indicate that a person who belongs here is: helpful, friendly, nice, responsible for others, able to work together on challenging work. For male faculty, there is an additional element of physics identity: They are aware of the challenges that women face in physics, and they see it as their responsibility to take those challenges seriously and take action on them. Interpersonal interactions in this setting are, in short, saturated with caring: Older, more experienced physicists (faculty; third and fourth year students) care for younger people (less experienced faculty; students; first and second year students). This care is sophisticated; it includes a willingness to take seriously the social location of others (men's acknowledgement that their experiences in physics are different from those of women's). Care is strongly associated with the feminine; thus, this caring environment is more welcoming to women students than the prototypical physics environment, where the dominant cultural theme is likely to be competition.[3]

4.6 Cultural Domain

To get at the cultural domain and what it can tell us about acceptable physics identity in this setting – acceptable ways of being a physics major – I directly asked both students and faculty what they think it means to be a good physics student. I also looked at the cultural implications of the physical spaces in physics – what messages the setting itself conveyed about who belongs in physics. Finally, I asked faculty what they think their department is doing to be a good place for women in physics, and some of their answers also helped me understand the cultural domain.

4.6.1 What do academic faculty members' words and actions convey about what's valued in this setting?

I asked all the physics faculty what they think it means to be a good physics student. Three of their responses were detailed and rich (more on the fourth in a moment). All three of them brought out these themes: Good physics students are *curious, interested and engaged*; they *think critically about concepts and abstractions*; and they should be prepared, through their physics major, to live *productive, happy, good lives,* but they *don't have to pursue a PhD.* One faculty member told me that to be a good physics student means "to be curious and work hard. Curious in that you ask questions – you have to have enough confidence to ask questions, and realize – it does take some confidence to realize you can ask questions and it's not a bad reflection of you." Another said she's also "looking for enjoyment – they say 'this is cool' at one point during the semester. Otherwise it seems like you're setting yourself up for four years of pain!" A third told me "We want everyone to be good physics students, but they don't have to all be great physics students. They have to be successful at acquiring various useful skills. They're not all going to be physicists, and we want them to be productive and happy." Two faculty members also said that good physics students can *work together* and one said they should be able to *work hard.* One told me that he would judge someone to be a good physics student "if the

[3] A student who had transferred from a large research intensive university talked explicitly about the non-competitive atmosphere in this physics setting: "The relationship you have with teachers here is worth noting. They want physics majors, and the teachers want you to learn and understand the stuff. At [my previous institution] you're competing against other students for the curve – you do really well and you hope other people don't understand it as well as you so the curve is in your favor. Teachers there, they know a third of you are going to fail no matter what. Here they really want you to understand it."

student is turning in work, and participating in the group work, and working at it, and willing to make multiple attempts."

The fourth faculty member gave me an answer in which he didn't discuss the elements of physics identity he is looking for in his current setting; he did, however, explicitly reject an older form of physics identity. "What does it mean for someone to be a good physics student? I don't know! I would have given you a very different answer when I started my career. The answer back then was 'someone who's a lot like me' – does really well in this or her physics classes, goes on to get a PhD in grad school, and then can't get a job when they get out because there are no jobs for PhD physicists now. Right now I'm not sure I know the answer to that question very well. This may sound weird, but I'm not sure I really want to have in my mind a vision of what I would consider to be a good physics student, because any concrete model that I have in my mind, a person I consider to be a good physics student, is going to be exclusionary."

4.6.2 What do students' words and actions convey about what's valued in the setting?

These cultural themes were echoed in other data sources. The idea that good physics students are *curious* and *enjoy abstractions* came out in the focus group I held with third and fourth year physics students, during which I asked students why they chose to major in physics. One person answered that they "like knowing how things work," and 12 other people (virtually the entire group) indicated that this was one of their primary motivations as well. During an interview, a fourth year student told me that "My parents, physics people here – just make fun of me because all I talk about is [deep significant voice] space. But at least people appreciate it! That is part of why I made friends with physics majors." The students in the focus group also emphasized that majoring in physics entails *long, hard work* in the physics building:

> Me: "What characterizes majoring in physics?"
> Students: "The physics building."
> "We live here."
> "If I wasn't a commuter I'd be there 24-7."
> "Spending more hours in the physics building outside class than you do sleeping."
> [Someone suggests a 3rd story of the physics building so people could sleep there.]
> "We could save so much money we pay for residence if we could just live here."
> "I slept on a table."
> "[Another student] waking me up at 2 in the morning on a professor's couch, asking if he could have a turn."
> "Waking up just before dawn, being relieved at least the sun wasn't up yet."
> "The sign in the lounge asking people to be out before the cleaning staff was there."
> "Is there life beyond this building? That's the question."

4.6.3 What does the space itself convey about what's valued in the setting?

Physical and virtual spaces in physics underscore the messages that in this department, physics is about *working together* and that physics majors *don't have to pursue a PhD*; many career routes are celebrated. The building where physics classes and labs meet and where the physicists have their offices has several open spaces where the physics majors are encouraged to work together, including a lounge (formerly a very small seminar room) for the physics students. There are posters on the wall promoting high school teaching as a career and indicating that physics majors score high on the LSAT and MCAT. The department website lists careers that physics grads have gone on to pursue, including, as one faculty member put it, "med school, geology, earthquake, undersea things" (and I must say that a career in undersea things sounds pretty great). The latter message is also underscored by an annual event: A career panel of physics alums. A faculty member told me that their intent with this panel is to convey "you don't have to go to grad school to be a physics major. There are many other things you can do, and we're excited about them, they're cool jobs, they're a good way to spend your time, to spend your career. It isn't 'you have to look exactly like us to be a physicist.' There are many options."

One faculty member told me about a concern he has about a poster of Einstein displayed outside his office. "I sometimes feel guilty having this poster here, because it's perpetuating the myth of the lone genius, and this epiphany takes place and changes the world. That's not really how it's done. It's incremental, it's social and interactive, and I actually find that empowering. I'm not Einstein and I can still do physics!"

4.6.4 How do the things that are valued align with larger cultural beliefs about race and gender? Conform to or contest common patterns within physics settings? In society?

The cultural domain, like the interpersonal domain, indicates that assuming a physics identity in this department involves being collaborative and working hard; the physics identity is expanded in the cultural domain to indicate that desirable physics students are curious and engaged, and enjoy thinking abstractly. However, good physics students are not expected to go on to become research scientists; there is support in the cultural domain for students to pursue a wide range of careers. Again this promotes a physics identity which is more comfortable for women – especially women of color – to step into, because it emphasizes that success in physics is about interest (which all the women in this study had in abundance) and hard work. This contradicts the idea, common both in physics settings and in US society, that success in physics comes from natural ability; that successful physicists are (White, male) geniuses who work in isolation.

4.7 Structural Domain

4.7.1 What are the policies in classrooms, labs and other physics spaces?

This environment – in which students support one another and faculty support students – is created and maintained at the structural level. Several years before I carried out this study, the academic faculty in the department began adopting the Student Centered Active Learning Environment for Undergraduate Programs (SCALE-UP) protocols (Beichner 2008). They also strive to use the best practices outlined in the 2003 report from the American Association of Physics Teachers (commonly called the Spin-Up report) on the characteristics that differentiated thriving physics departments from those that were declining in size or not graduating many students (Hilborn et al. 2003). These characteristics include:

- "A widespread attitude among the faculty that the department has the primary responsibility for maintaining or improving the undergraduate program. That is, rather than complain about the lack of students, money, space, and administrative support, the department initiated reform efforts in areas that it identified as most in need of change.
- "A challenging, but supportive and encouraging undergraduate program that includes a well-developed curriculum, advising and mentoring, an undergraduate research participation program, and many opportunities for informal student-faculty interactions, enhanced by a strong sense of community among the students and faculty.
- "Strong and sustained leadership within the department and a clear sense of the mission of its undergraduate program.
- "A strong disposition toward continuous evaluation of and experimentation with the undergraduate program" (Hilborn et al. 2003, p. vi).

These features permeate the physics department; both faculty and students allude to these practices, and I saw them enacted during my participant observation in physics settings. As one faculty member told me, "we now teach the introductory course using SCALE-UP methods. … *Group work* and lab work – or lab exercises which take the place of lab – are part of the course work. Group work is built in, lectures are very strongly de-emphasized in favor of a lot of group work, a lot of *interaction between faculty and students*, fairly short lab exercises which are pertinent to the material being taught rather than having separate lab sessions which could be a week behind or a week ahead of the class material at the time." Students report the same thing; one woman told me "I like that the classes aren't lectures.

Most of the teachers will lecture for a half hour, 45 minutes, and then at some point in there ends up being some partner work or someone goes up to the board and works through a problem. So it's very interactive, instead of just being talked at. It's more a conversation with everybody in the class and the professor than information being thrown at you, because that's not helpful. I think that's a big thing to me as to why I enjoy it and I have learned so much from it."

The Spin-Up protocols aren't the only use the faculty make of research; they *use research to guide all their practices* (and this is in fact how I initially gained their trust to be allowed such free access to their classes and students – for years before I carried out this study, the physics faculty had used me as a sounding-board as they explored educational research, trusting my expertise in education coupled with my previous career as a high school physics teacher). Another member of the physics faculty told me "We try to pay attention to the things studies show is useful, instead of what we feel is useful. The style of teaching that we do – the interactive style" was adopted because of its research grounding. The physics faculty not only make use of physics education research, they actively contribute to it, including publications in *The Physics Teacher* and involvement in the local and national branches of the American Association of Physics Teachers. One of the principal manifestations of this commitment to research is through the *constant department-wide use of both formative and summative assessment*; the faculty constantly check to see what students are learning during each class session and they also make use of the Force Concept Inventory[4] (FCI; Savinainen and Scott 2002) and its sister assessments to compare their students' conceptual growth across the semester both to students from other institutions and to their own students during other semesters. A faculty member told me that "our data match national statistics – faculty who use interactive methods, and strongly use interactive methods, and group work in class, and feedback between students and faculty members have higher FCI gains than faculty who don't." Another faculty member told me that "we're pretty well-off with our evaluation stuff, our assessment things, so we try to stay on top of that, we try to stay in front of it because a) we're going to have to do it anyway, b) it's better to choose an assessment that actually assesses things you want to know. Then we can do things that actually make a difference, as opposed to just guessing."

Two of the faculty members use clickers to monitor student understanding during class (they project a conceptual question with several possible answers on a

[4] The Force Concept Inventory is a multiple choice test that assesses a student's conceptual understanding of force and Newton's Laws. When the FCI was first introduced, physics teachers and professors across the US (and I count myself in this number) were horrified to discover that some of their students could solve physics problems proficiently while still holding utterly wrong beliefs about how the world works – they had learned to solve problems but hadn't actually learned physics. This physics department uses the FCI not to assess students but to measure their own effectiveness – they administer it at the beginning of the introductory course and again at the end to see whether their teaching approaches were effective in helping students identify and correct their misconceptions about force. Since the development of the FCI, a number of similar tests have come into use; see for instance https://www2.ph.ed.ac.uk/AardvarkDeployments/Public/60100/views/files/ConceptualTests/Deployments/ConceptualTests/deploymentframeset.html

screen; students use hand-held devices to "click" the answer they think is correct; faculty monitor the answers on their laptops and modify their instruction accordingly). A third uses the old-school method of having students simply hold up cards to indicate which answer they prefer. A student describes this process: "We have these cards for answering the multiple choice questions in our groups, and it's for him to see where everyone is. We'll like answer the question, and maybe everyone's cards will have a different letter, so we discuss why we chose which one."

Faculty members expect students to work in groups, but they don't leave this process to students. They *teach students how to work effectively in groups*, and grade them on their success. "We talk about – instead of just modeling it – we talk about 'here are the things you need to do to work in a group, here it's important to be explaining what you're doing.' We provide feedback as well – things like group work, where we're trying this CATME system [a set of electronic tools that support group work], stuff like that – frequent feedback, even on the more concrete physics skills – trying to provide frequent feedback starting out in low-stakes situations, then eventually there has to be something at stake to incentivize it." Faculty also intervene when students are failing to work collaboratively in their groups; I will talk more about this in the disciplining section, below.

Finally, the academic faculty in the department systematically convey that physics is learned through *hard work and practice, not natural ability* (Dweck 2006). A faculty member told me how they do this: "We explicitly mention [theories of growth mindset], we're emphasizing the discussion and exploring, making it safe to make mistakes – doing conceptual questions with clickers and having students have to talk about it – a low-stakes environment, they see that there are lots of other people who don't fully grasp it, and through working at it they collectively develop their understanding." Another faculty member pulled together all these themes: "Active learning – it gets everybody involved, lets you know if there are problems and it kind of makes everybody act on the knowledge—'maybe I don't know this.' The message is '*you can be wrong and still be a physicist.*'" I witnessed this in class; for instance, in a 400-level class where students were busily working on the board, a student put up an answer, returned to his seat, then announced" I think I need to change it." The physicist teaching the class responded "you're allowed to modify!" and he got up and changed his answer. Then after a few minutes, as he listened to students discuss other problems, he jumped up and changed it yet again; clearly there was no penalty in this setting for taking some time to understand a challenging physics concept.

During the focus group I conducted with third and fourth year physics majors, a student who transferred to this institution after 2 years at a large research university contrasted the experiences: "There's a lot of interaction in class that you won't get at a place like [previous institution]. At [that university] you can show up and scribble stuff down and you leave. Here's an example. [My current professor] – she's like painting a picture, and she wants you to fill in the trees, make a metaphor. She was doing these differential equations, and like 'what are the boundary conditions?' And we all know it, but we might be wrong, so we wait like ten seconds, then we say 'x=0 and x=L' and she's like 'that's right!' … You don't even have to show up to class at [previous institution]. 300 person lectures. It was really easy – no accountability until test time rolled around. Here you feel awkward if you aren't in class."

4.7.2 Do the policies in place serve to perpetuate or challenge under-representation in physics?

One faculty member told me that teaching in the department is organized around these themes – especially around the idea that physics is done in collaboration and involves hard work and practice, not natural genius – explicitly because they want a more diverse student body. When I asked why he thought that the women physics majors were reporting such good experiences, he told me "you can talk about some of the things we don't do from the traditional methods that were problematic – 'weed them out, produce someone that looks like me, another physicist' approach is problematic if you're already starting with a bunch of white males as the professors. Also the whole idea of – 'you've got to come in with some innate talent' – 'the lone genius' – 'let's revere Einstein and Feynman' – that leads to people selecting people who look like them. There was a study in *Science*, I've referred to it like 3 times this week – where they looked at a bunch of different disciplines – not just in the sciences; economics, philosophy, psychology, English. One of the things that correlated with underrepresentation of women was disciplines in which more people subscribed to the genius model – fixed intelligence vs a growth mindset kind of thing."[5]

So teaching practices in the structural domain are crucial for creating a physics identity that women of color can step into without much difficulty. But it's the practices of the academic faculty in the disciplining domain that make it clear that this doesn't just happen by itself; it requires constant maintenance.

4.8 Disciplining Domain

4.8.1 What kinds of student behaviors do academic faculty correct?

I collected a number of instances of physics faculty members correcting students' behavior. The same theme emerged in this domain as the interpersonal, cultural, and structural: The importance of collaborative group work. In the disciplining domain,

[5]He is referring to the study by Leslie, Cimpian, Meyer and Freeland (Leslie et al. 2015) that showed that the under-representation not only of women but also of African Americans was tightly correlated to the perception by people in a particular field of the importance of natural ability to success in that field.

however, I was able to see that faculty mean what they say and take responsibility for making it happen. I saw several instances of faculty *reprimanding students who failed to work equitably in groups*, and faculty told me about other instances. For example, during the first day of an introductory class, a member of the physics faculty was explicitly telling students about how to work in groups to solve problems at the whiteboard. He offered a tip on how to make this happen: On different problems, have different people hold the marker. Despite this suggestion, one student continued to maintain control of the marker, and said that it would be more efficient if he and his group members just divided up the problems. In response, the faculty member said that *the goal of group work is not efficiency but that everyone learns*. In a similar example, a faculty member told me about working with a student who was dominating group work during a lab. He was controlling all the materials, so she told him he had to let other people have a chance, at which point he backed up and stood far away from his group. She told him he didn't have to stand so far away and that he was either dominating the group or not participating. According to this faculty member, she said "You can't only participate when you're building, that's not OK. It can't be 'I'm either in charge or I'm out of here, guys.'"

Note that in these examples, faculty were reacting to students who were trying to assert the prototypical physics student identity I laid out in Table 4.1: that good physics students are the ones who can get the answers on their own; that physics is about individuals getting the right answer (ideally faster than others), not about working collaboratively or ensuring that everyone learns. This prototypical identity is aligned with masculine norms and domains – competition, individuality, winning. For male students, adopting the physics identity which is required in this setting involves a loss; it involves giving up a sense of their own superiority which they have likely used in the past to bolster their egos, and it involves developing some skills that are looked on as feminine (concern for others; the ability to work collaboratively). One faculty member told me about dealing explicitly with issues of *gender and group work* when giving students feedback. She was dealing with a situation in which two male students were in a lab group with a woman, and they almost entirely excluded her from participation (I was present in this lab and witnessed this). After the lab ended, the faculty member talked with all three of them about it. The excluded student said she was OK with what happened, but the faculty member told me "it wasn't OK to me. It didn't really matter what [the woman student] thought. She can have her own subjective opinion on it. My opinion on it matters more." She suggested to all three group members that "they pay attention to the way they interacted" – men with women, women with men – "to see if women would ever say no. Because it really surprised when I started to pay attention. I was surprised how often I would not say no, but instead come up with a better idea, and argue about it. I didn't realize you could say no."

4.8.2 What kinds of student behaviors do students believe academic faculty correct?

The women physics majors believe that *their faculty would protect them from sexism*. While I was carrying out this study, I was invited to present at a Conference for Undergraduate Women in Physics.[6] I invited all the women physics majors and both women physicists to a session where I practiced my presentation. Almost all the students and both faculty members attended. The presentation included some experiences women of color have had in physics (experiences collected by me and also by Mia Ong and Apriel Hodari). One of the stories I presented had actually happened to one of the women in the room while she was doing summer research at another institution (called a research experience for undergraduates, an "REU"), and she had given me permission to include it (without revealing that it was her story). I presented the story on a slide, and everyone read it silently:

> I was both the only girl and the only undergrad in the entire lab. I didn't actually work with the man who hired me, I worked with a graduate student. I was working with this one volatile chemical to try and density match things, and another person walked into the lab and bumped me while I was pipetting the liquid, and it spilled onto the lab table. I moved one of the hoods over it and someone else walked into the lab, commented on the fact that it smelled, and my mentor laughed and said 'can you guess who spilled it?' and they all looked at me, and they all started laughing, and I was the only girl in the lab at that point, and they all continued to laugh, and I just kind of stood there awkwardly, and the grad student said "how does it feel to have the boys club laughing at you?"

This prompted some discussion, with one student saying she wouldn't know what to say, she would just be silent if this happened to her. Then I posted a slide about what happened next:

> I just packed my things up and left that day. [For the rest of the summer] I didn't speak except when spoken to. I was too afraid to reinforce their idea that I was incapable and didn't belong there. I was afraid of making the necessary mistakes to succeed. I spent basically an entire month being silent in a lab and trying to just work 8 to 6 like I was supposed to, and just leave.

At this point, one of the faculty members, appalled, said "wait, this happened in an REU?" Throughout the rest of the workshop, the faculty member kept coming back to this story and talking about what a jerk the graduate student had been. When I asked the women in the room what could have been done to make this incident better, one of them said "having our professors!" – and, indeed, when I had asked the same question of the woman who originally told me the story, she told me she wasn't sure what would have happened had the same incident occurred at her undergraduate institution; she said "I would hope that someone would speak up and be

[6]The first Conference for Undergraduate Women in Physics took place at the University of Southern California in 2006. Since then the number of conference locations has grown annually; in 2018 there were 9 conferences held simultaneously at different locations throughout the US and, for the first time, one in Canada as well.

like 'that's wrong, you shouldn't say that to another person,' but I think probably generally there would be laughter that followed it." Although she wasn't sure how her peers would react, she was confident that at least one of the members of the physics faculty would have intervened: "I'm sure if I said something to [physicist], he would pull that person aside and have a conversation with them, because he is kind of like the dad of the physics department."

4.8.3 How do the student actions that academic faculty condemn or correct align with or challenge under-representation in physics?

These faculty members don't just talk about valuing diversity; they are willing to step in and insist on student behaviors that create a more welcoming environment. Women students believe that the members of the physics faculty will protect them from the most egregious behaviors of male students.

4.9 Discussion

In Table 4.4, below, I have summarized the components of physics identity that I was able to pinpoint by gathering data in the interpersonal, cultural, structural and disciplining domains of power. The component of student identity which emerged in every domain is that physics students work collaboratively together. The cultural domain added more components to what it means to be a physics student in this setting: Physics students are curious and engaged, and can think critically, but don't need to be on track to become research scientists; physics majors can aspire toward a variety of satisfying careers. The most important components of physics faculty identity emerged in the structural domain (and were reinforced in the disciplining domain): Physics faculty members use research-based teaching strategies including high levels of faculty-student interaction, collaborative group work, and extensive assessment of student learning. There are two more components of physics faculty identity that I think are particularly crucial for this setting: Physics faculty believe that success in physics is a result of hard work rather than natural ability, and male physics faculty take gender issues in physics seriously, rather than leaving equity issues to their female colleagues.

Table 4.4 Components of physics identity in this setting

Domain	Student identity	Academic faculty identity
Interpersonal	Physics students work together	Physics faculty are acessible to students ("nice", "helpful")
	Physics students help one another	Physics faculty are accessible to their colleagues ("super-generous")
		Physics faculty seek shared agreement with one another through discussion
		Male faculty take responsibility for gender issues
Cultural	Physics students are curious, interested and engaged	Physics faculty are proud of the range of jobs their alums work in
	Physics students think critically about concepts and abstractions	
	Physics students should be able to use their major to pursue productive, happy lives	
	Physics students don't need to be on the physics PhD track	
	Physics students work hard and spend a lot of time in the physics building	
	Physics students work together	
Structural	Physics students do group work in class	Physics faculty interact extensively with physics students in class
		Physics faculty use research to guide their practices
		Physics faculty use assessment to guide their teaching
		Physics faculty teach students how to work effectively in groups
		Physics faculty believe hard work, not natural ability, leads to success in physics
Disciplinary	Physics students should be able to work smoothly in groups, ensuring that all group members participate	Physics faculty take responsibility for making sure group work goes smoothly, by having difficult conversations with students who are not effective group members
	Physics students should be committed to the learning of all group members	Physics faculty protect students from racist, sexist microaggressions

In Table 4.5, below, I return to the description of prototypical physics departments that I included in Table 4.1, and compare prototypical departments to this department. This comparison makes it clear just how different the interpersonal, cultural, structural and disciplinary domains of power are in this setting than in many other physics departments.

Table 4.5 Comparing characteristics of domains of power in a prototypical physics department and the department under study

	Prototypical physics departments	This physics department
Interpersonal domain	Women, especially women of color, are isolated and experience microaggressions	Students are friendly and helpful; they work on problems together (even if they don't really like group work) and socialize together
Cultural domain	Physics is a competition; good physicists are natural geniuses who work in isolation	Physics is collaborative; success in physics results from hard work and practice; physicists can be wrong
Structural domain	Classes consist of faculty lecturing; high levels of student failure are tolerated	Faculty and students are highly interactive; students work collaboratively during class to learn physics and solve problems
Disciplinary domain	Faculty do not intervene in student-student interactions	When students contest this collaborative culture and attempt to assert a prototypical physics identity, faculty intervene

I analyzed this setting using an intersectional physics identity framework; however, my findings have for the most part been relevant to why this setting is comfortable for all women; my insights have not been particularly intersectional. The setting is marked by an emphasis on collaboration, caring and hard work, skills that I would argue are stereotyped as neutral-to-feminine – in sharp contrast with more prototypical physics settings, where the more stereotypically masculine skills of competition and natural ability are venerated. Thus, women in this setting are not under the stereotype threat that they can experience in other physics settings; they are not, by virtue of their gender alone, disadvantaged by having to disprove stereotypes about people-like-them. On the contrary, male students may be under stereotype threat, particularly in regards to caring. (One imagines, however, that the tremendous stereotype lift that men get by being in physics more than offsets this disadvantage).

The ways I have identified in which women of color would particularly benefit from this setting are a question of quantity more than quality. All women are isolated in most physics settings; women of color are isolated even more. All women are under stereotype threat in physics; Black women and Latinas are doubly stigmatized as not smart enough, based on not just gender but racial stereotypes as well.

Because there are so very few women of color majoring in physics in the United States, it is difficult to study how, for instance, Black women's experiences might differ from those of Latinas or White women or Asian women.[7] I'm not sure that the questions I proposed in my intersectional identity framework are sufficient, either. The framework was powerful in helping pinpoint how this particular setting differs from prototypical physics departments, and those differences made it very clear why the physics student identity and the physics faculty identity are both manageable for women to assume in this setting. These identities downplay the more

[7] Of the 6283 people who received bachelor's degrees in physics in 2014 across the entire United States, for instance, 81 were Latinas, 52 were African American women and 5 were America Indian women (National Science Foundation and National Center for Science and Engineering Statistics 2017).

masculine characteristics that prototypical physics settings emphasize; instead they rely on neutral-to-feminine skills and characteristics, like curiosity, collaboration, hard work, caring, and (for faculty) embracing best teaching practices.

I want to end this chapter with another event from my practice presentation. After I had finished going through all my slides, one of the faculty members said that they stirred up a lot of feelings for her. I asked her what she'd done when her belonging in physics had been questioned, and she said she'd reminded herself that she has a PhD, she knows what she's talking about, she has plenty of objective evidence that she is competent in physics. Then the other faculty member said that what she has done is "out-compete them." When she was in the middle of her physics major (near the end of her second year) her physics instructor leaked everyone's grades, and it turned out that she was a whole letter grade ahead of everyone else. From that point on, she said, no-one would work with her and she had no friends in her major for the rest of her time as an undergraduate. After she told this story, there was a silence that went on and on; it seemed to me that the other women in the room were feeling for her, imagining what her isolation had been like, realizing the cost she had paid so that she could create a very different place for them to do physics.

References

Anzaldúa, G. (1999). *Borderlands/La frontera* (2nd ed.). San Francisco: Aunt Lute.

Barthelemy, R. S., McCormick, M., & Henderson, C. (2016). Gender discrimination in physics and astronomy: Graduate student experiences of sexism and gender microaggressions. *Physical Review Physics Education Research, 12*(2), 020119. https://doi.org/10.1103/PhysRevPhysEducRes.12.020119.

Beichner, R. (2008). *SCALE-UP: Student-centered active learning environment for undergraduate programs*. Retrieved from http://scaleup.ncsu.edu/

Carlone, H., & Johnson, A. (2007). Understanding the science experiences of successful women of color: Science identity as an analytic lens. *Journal of Research in Science Teaching, 44*(8), 1187–1218.

Collins, P. H. (2000). *Black feminist thought* (2nd ed.). New York: Routledge.

Collins, P. H. (2009). *Another kind of public education: Race, schools, the media, and democratic possibilities*. Boston: Beacon Press.

Collins, P. H. (2015). Intersectionality's definitional dilemmas. *Annual Review of Sociology, 41*, 1–20.

Collins, P. H., & Bilge, S. (2016). *Intersectionality*. Chicester: John Wiley & Sons.

Crenshaw, K. (1989). Demarginalizing the intersection of race and sex: A black feminist critique of antidiscrimination doctrine, feminist theory and antiracist politics. *The University of Chicago Legal Forum, 140*, 139.

Crenshaw, K. (1991). Mapping the margins: Intersectionality, identity politics, and violence against women of color. *Stanford Law Review, 43*(6), 1241–1299.

Dweck, C. (2006). *Mindset: The new psychology of success*. New York: Random House.

Fullilove, R., & Treisman, U. (1990). Mathematics achievement among African American undergraduates at the University of California, Berkeley: An evaluation of the Mathematics Workshop Program. *The Journal of Negro Education, 59*(3), 463–478.

Ginther, D. K., Schaffer, W. T., Schnell, J., Masimore, B., Liu, F., Haak, L. L., & Kington, R. (2011). Race, ethnicity, and NIH research awards. *Science, 333*(6045), 1015–1019. https://doi.org/10.1126/science.1196783.

Gonsalves, A. J., Danielsson, A., & Pettersson, H. (2016). Masculinities and experimental practices in physics: The view from three case studies. *Physical Review Physics Education Research, 12*(2), 020120. https://doi.org/10.1103/PhysRevPhysEducRes.12.020120.

Guy-Sheftall, B. (1995). *Words of fire: An anthology of African-American feminist thought.* New York: The New Press.

Hilborn, R. C., Howes, R. H., & Krane, K. S. (2003). *Strategic programs for innovations in undergraduate physics: Project report.* College Park: American Association of Physics Teachers (NJ1).

Johnson, A. (2007). Unintended consequences: How science professors discourage women of color. *Science Education, 91*, 805–821.

Johnson, A., Brown, J., Carlone, H., & Cuevas, A. (2011). Authoring identity amidst the treacherous terrain of science: A multiracial feminist examination of the journeys of three women of color in science. *Journal of Research in Science Teaching, 48*(4), 339–366.

Johnson, A., Ong, M., Ko, L., Smith, J., & Hodari, A. (2017). Common challenges faced by women of color in physics, and actions faculty can take to minimize those challenges. *The Physics Teacher, 55*, 356–360.

Kachchaf, R., Ko, L., Hodari, A., & Ong, M. (2015). Career–life balance for women of color: Experiences in science and engineering academia. *Journal of Diversity in Higher Education, 8*(3), 175–191.

Ko, L. T., Kachchaf, R. R., Hodari, A. K., & Ong, M. (2014). Agency of women of color in physics and astronomy: Strategies for persistence and success. *Journal of Women and Minorities in Science and Engineering, 20*(2), 171–195.

Leslie, S.-J., Cimpian, A., Meyer, M., & Freeland, E. (2015). Expectations of brilliance underlie gender distributions across academic disciplines. *Science, 347*(6219), 262–265.

Moraga, C., & Anzaldúa, G. (Eds.). (1981). *This bridge called my back: Writings by radical women of color.* California: University of California Press.

Moss-Racusin, C. A., Dovidio, J. F., Brescoll, V. L., Graham, M. J., & Handelsman, J. (2012). Science faculty's subtle gender biases favor male students. *Proceedings of the National Academy of Sciences, 109*(41), 16474–16479.

National Science Foundation. (2016). *Science and engineering indicators.* Retrieved May 25, 2017 https://www.nsf.gov/statistics/2016/nsb20161/#/report

National Science Foundation., & National Center for Science and Engineering Statistics. (2017). *Women, minorities, and persons with disabilities in science and engineering: 2017. Special report NSF 17-310.* www.nsf.gov/statistics/wmpd/

Office for Civil Rights. (2014). *Civil rights data collection: Data snapshot (College and career readiness).* Retrieved from Washington, DC

Ong, M. (2005). Body projects of young women of color in physics: Intersections of gender, race, and science. *Social Problems, 52*(4), 593–617.

Savinainen, A., & Scott, P. (2002). The force concept inventory: A tool for monitoring student learning. *Physics Education, 37*(1), 45–52.

Seymour, E., & Hewitt, N. (1997). *Talking about leaving.* Boulder: Westview Press.

Smith, B. (1983). *Home girls: A black feminist anthology.* New Brunswick: Rutgers University Press.

Spradley, J. (1979). *The ethnographic interview.* New York: Hold, Rinehart and Winston.

Spradley, J. (1980). *Participant observation.* New York: Holt, Rinehart and Winston.

Traxler, A. L., Cid, X. C., Blue, J., & Barthelemy, R. (2016). Enriching gender in physics education research: A binary past and a complex future. *Physical Review Physics Education Research, 12*(2), 020114. https://doi.org/10.1103/PhysRevPhysEducRes.12.020114.

Treisman, U. (1992). Studying students studying calculus: A look at the lives of minority mathematics students in college. *The College Mathematics Journal, 23*(5), 362–372.

Zinn, M. B., & Dill, B. T. (1996). Theorizing difference from multiracial feminism. *Feminist Studies, 22*(2), 321–331.

Chapter 5
Urban College Students Negotiate Their Identities to Dis/Connect with Notions of Physics

Diane Crenshaw Jammula and Felicia Moore Mensah

5.1 Introduction

Pioneers in the field of physics education research (Hestenes 1987; Karplus and Brunschwig 1969; McDermott 1995) sought ways to teach physics such that students build deep conceptual understandings of physics phenomena and develop expert-like views of the epistemology and nature of physics. From interactive engagement in physics learning, such as conducting experiments, engaging in argumentation, and solving problems in groups, students achieve almost twice as much in learning gains as compared to the more traditional lecture-style of instruction (Von Korff et al. 2016). However, gender, race, and ethnic achievement gaps persist even with the use of interactive instruction, though they do not widen (Brewe et al. 2010; Pollock et al. 2007). To explain these continued discrepancies, Kost-Smith, Pollock, and Finkelstein (2010) suggest a "smog of bias…that surrounds us and that we constantly breathe in, though at times we may be unaware that it even exists" (p. 15). Like smog, gender bias is omnipresent and toxic, yet sometimes invisible. However, the harmful impact of smog on public health may be more widely accepted than the impact of gender bias on women's wellbeing. The "smog of bias" referred to by Kost-Smith et al. (2010) suggests an unconscious and pervasive privileging and discrimination in the physics classroom.

Gwyneth Hughes (2001) describes that science is taught as "a body of authoritative, incontestable knowledge which is abstracted from social activity to maintain a high level of difficulty and status" (p. 276). She argues that physics is the

D. C. Jammula (✉)
Rutgers University, Newark, NJ, USA
e-mail: diane.jammula@rutgers.edu

F. M. Mensah
Teachers College, Columbia University, New York, NY, USA
e-mail: fm2140@tc.columbia.edu

A. J. Gonsalves, A. T. Danielsson (eds.), *Physics Education and Gender*, Cultural Studies of Science Education 19,
https://doi.org/10.1007/978-3-030-41933-2_5

most positivist and elitist of the sciences. Consequently, few students can connect to this exclusionary perception of the nature of physics. Yet, dis/engagement with the subject depends on more than one's gender. Some females prefer abstract, rational thinking while some males are deterred by it. Rather than essentialize "how females do physics" and "how males do physics," Hughes shows how students negotiate their multiple subjectivities, including gender, race, ethnicity, and class, to align with or disengage from their views of science. Students may see physics as abstract, rational, and elite, which may or may not match views of themselves. This notion of physics as abstract, rational, and elite may contribute to the lack of diversity in participation and achievement in physics where students feel they have to perform a certain way to have an identity in physics.

5.2 Gender Subjectivities and Performance

Judith Butler's (1988) notion of gender performance argues that one expresses their subjectivities in discourse, or the way they communicate themselves to others, including speech, bodily movements, and style. Children learn gender (and race, ethnicity, and class) performance by watching and participating in the social world. They interact with men and women in their lives, watch men and women on TV, and are disciplined by individuals and institutions to be boys or girls. This viewpoint might explain why in a survey of 437 sixth-grade students, boys had more experience with tools, electronics, and simple machines, while girls had more experience knitting, cooking, and gardening (Jones et al. 2000). Women are not biologically determined to be creative, caring, and charming, but they may be disciplined as females to adopt these characteristics. Invoking the category "women" assumes a well-defined group of human beings. It suggests that "women" share commonalities, which are different from "men." Yet, not all women, or human beings, are one way or the other. There may be greater variation within the categories of men and women than between them (Epstein 1988).

Moreover, students negotiate their subjectivities to dis/connect with notions of physics. When negotiating subjectivities, one draws from their subject positions to engage with their environment. For example, Hughes (2001) writes about a female student of color who does not ascribe to traditional femininity. The student rejects the notion of scientific knowledge as coming from authority. Instead, she embraces a constructivist nature of science that aligns with her gender and ethnic subjectivities. She develops a positive science identity. Carlone and Johnson (2007) propose a model for science identity that includes competence, performance, and recognition. While women and people of color may not be recognized by others as "science people," they may see themselves as science people because of their love of science or agency in redefining science (Rosa and Mensah 2016). Thus, populations marginalized in the science classroom may still form positive science identities by connecting with their views of the subject matter.

This study conceptualizes gender as a spectrum of performances associated with femininity and masculinity in the physics classroom. In the U.S. in the twenty-first century, collaborative, emotional, and caring are associated with femininity and competitive, rational, and self-interest are associated with masculinity, to name a few. These associations are socially constructed; they depend on how a society defines what it means to be a woman or man, and therefore depend on time and place. The research question addressed in this study is: how do students negotiate their subjectivities to dis/connect with notions of physics?

5.3 Ethnographic Methods

Ethnographic methods (Emerson et al. 2011) were implemented to capture the classroom culture in the physics classroom and to understand within this setting how students negotiate and dis/connect with notions of physics. From the field of anthropology, ethnography is used to develop understandings of different cultures. The physics classroom has a culture, including norms, values, beliefs, rituals, roles, and power structures. This culture is apparent in the way participants engage with each other in activities within the classroom space. Data such as fieldnotes, audio transcripts, and classroom artifacts is collected to form a "thick description" (Geertz 2008) of the classroom culture. This methodology was implemented to see how students enact their subjectivities and how such actions work to construct notions of physics.

The study setting was an urban public college in New York City. The 6-year graduation rate at the college is 42%. The demographic make-up of the college was 40% Latinx, 25% African American, 10% Asian American, and 25% White, which is representative of the typical course enrollment. The physics course for this study was an interactive algebra-based introductory physics course that was taught by the first author. Instead of a traditional lecture hall, students sat at lab tables in groups of four. Each student kept journals and wrote notes and reflections as they related to activities in the course. They designed and conducted experiments to build models of physical phenomena, including acceleration, force, energy, and momentum. After each experiment the student groups presented their findings and engaged in argumentation to reach consensus about tenets of the phenomenon. They then applied the model of the phenomenon to different scenarios and problem solving. Once limitations of the model are reached (e.g. the model for constant velocity cannot be applied to objects that speed up, slow down, or change direction), students engage in a new cycle of model development and deployment.

There were 23 students enrolled in the course, comprising 7 female students and 16 male students. Over half of the class (13 students) will be the first in their family to graduate college. Students self-identified among a range of ethnic and/or racial backgrounds. Most students grew up in the U.S. and 3 students immigrated to the US as teenagers. The students in this course were non-science majors fulfilling a science requirement. Most were freshmen and sophomores, with a few juniors and seniors. The students ranged in age from 18 through mid-twenties. Upon completion of the course and after grades were submitted, 22/23 students consented to this study. However, for this study, five participants were purposefully selected because their journal entries and actions illustrated three different ways students dis/connect with physics. Three males (Ivan, Louis, and Greg) engage stereotypically; one female (Naira) performs less consistently; and one male (Sameer) excels in problem solving but does not ascribe to the dominant discourse of the physics classroom. (All names are pseudonyms). These five participants engage with each other in a problem presentation that shows the dynamics and outcomes of the enactment of their subjectivities.

5.4 Data Collection and Analysis

In the physics course, there were many artifacts that were collected and served as data for this study. Specifically, the researchers kept field notes of the classroom and interactions between students as ethnographic notes. The students wrote in journals each week about classroom happenings and their reactions, reflections, and feelings. This data source was significant as a personal document where participants shared their thinking about physics and how they dis/connected with physics. The journals from the consenting students were collected, and one large document (i.e., Journal File) was created and used for data analysis.

For data analysis, the Journal File document was analyzed inductively as a process for qualitative data analysis (Merriam and Tisdell 2015). We started with open coding (Emerson et al. 2011) and discourse analysis (Wood and Kroger 2000) to examine the journal entries and field notes. In the process of analyzing the journal entries, we read for moments where students discussed their subjectivities and views of physics. In open coding, many possible ideas were explored in making sense of the data as we coded the content of the journals. Phrases in journal entries were marked with short descriptors such as "experience," "view of science," "gender." We reread the coded journals and wrote notes and memos as the first level of understanding the participants' entries. We saw how the experiences, ideas, and dispositions of Ivan, Louis, and Greg shared commonalities. We pieced together memos from our analysis of their journals to form a narrative. The connection of middle-class white masculinity with conventional physics rang true in each of their stories. They also performed in similar ways in the physics classroom. No other participant wrote or acted like them, except on occasion Naira. The inconsistency of her actions caused us to select her journal data for further analysis and interpretation. Of all students, Sameer sharply contrasted with Ivan, Louis, Greg, and Naira. His journal was also selected for deeper analysis.

In the process of analyzing the field notes, we looked for moments where gender mattered. In other words, we analyzed moments when participants performed in ways associated with a particular gender in how they engaged in physics. For example, we wrote observational field notes when Greg controlled the computer in his group with two female classmates of color. We noted that as a moment where gender mattered. We were able to connect the fieldnotes to the journal entries because Greg also discussed this moment in his journal. Interpretations of fieldnote excerpts were elaborated in analytical memos.

Field note memos were integrated with journal narratives to deepen, elaborate, and develop arguments toward the development of themes that were coming from the analysis process. For example, the problem presentation was a classroom occurrence where gender mattered, and we tied together the narratives of the five selected participants in a meaningful way. As we worked with the data, we assigned pseudonyms and recorded the date of each entry to show when the entry was written (at the beginning, middle, or end of semester). This showed the persistence or changes of ideas over time. We maintained the participants' exact words, and punctuation and spelling edits were made only when necessary to make the excerpt more reader-friendly. The terms "masculinity" & "femininity," "white," and "middle-class" were used as concise labels for role behaviors that are commonly associated with either men or women, race or class designations, respectively. These categorizations do not represent innate or biologically determined categories of behavior, but they are culturally associated with particular gender, race, and class.

As researchers, we shared the process of analysis by reading excerpts from the journal entries and shaping them into stories that could be told. We served as peer reviewers for each other (Guba and Lincoln 1989) during the process to ensure both rigor and negation of meaning. We sought to highlight variation in how the participants negotiate their subjectivities to dis/connect with notions of physics. In this process, our goal is to bring forth a more comprehensive explanatory mechanisms for dis/engagement with physics, specifically in participants presenting the solution to a physics problem. We highlight below three themes as they relate to negotiating and disconnecting with notions of physics as middle-class white masculinity aligns with conventional notions of physics; as subjectivities are not essential; and as subjectivities are dynamic and gendered performances confer status.

5.5 Negotiating Subjectivities to Dis/Connect with Notions of Physics

This study focuses on one activity common to both traditional and active physics classrooms: solving a textbook physics problem. Findings show that subjectivities are intersectional, nonessential, and dynamic, and they reveal variations within the categories of gender, race, and class for the five participants. The participants draw from these subjectivities to connect with or turn away from their views of the discipline. The performances of subjectivities in the classroom activity of doing physics problems construct a general notion of physics as elite, authoritative, and rational. In this construct, *middle-class white masculinity* is privileged, and underrepresented voices are silenced, thereby reproducing hierarchies and discrepancies in physics education.

5.5.1 Middle-Class White Masculinity Aligns with Conventional Notions of Physics

Ivan, Louis, and Greg embrace the interests, performances, and discourses associated with middle-class white masculinity. These three participants appear as white males; Ivan is from Eastern Europe, Louis is Spanish (from Spain, as he indicated) and Caucasian, and Greg is white (he did not indicate ethnic background). They participate often in class and dominate whole-class discussions and small-group work. They speak with confidence and enjoy competition. These performances of dominance and entitlement align with conventional notions of physics as authoritative, elite, and masculine. The subjectivities of Ivan, Louis, and Greg overlap with a conventional notion of physics and they have positive physics identities. For example, Ivan describes his extensive experience and positive view of physics. He writes:

> In the past I have studied physics and I find it pretty interesting. It's one of my favorite subjects. I finished my high school in my country and there, physics was a subject that we had everyday. I have studied physics for 6 years, four in high school and two in college. (Ivan, 1/30)

Ivan's Eastern European background allowed him much experience in physics. In New York City public schools, physics is not a required course and only 20% of students take physics (Kelly and Sheppard 2009). In the following excerpt, Ivan explains his early like and understanding of physics:

> When I was in high school the first three years I never liked physics. Actually, I hated it. I rarely studied it. But on my senior year occasionally I used to read the book. At times I could understand it and to me it would make sense. I think from that year I really started to like it. I really like physics now. The reason for taking this course was that I somehow, I had missed that excitement when you get something right. (Ivan, 3/13)

Ivan constructs physics as authoritative. Physics knowledge comes from the textbook and success is defined by right answers and good grades. When Ivan is able to master physics content as dictated by the textbook and physics problems, he develops a positive physics identity. He describes a paradox that "you cannot learn science if you don't understand it," which highlights the exclusivity of the discipline and absence of entry points. Ivan's interest in physics stems from his feeling of success as determined by traditional indicators and not by curiosity about the physical world, testing ideas, or collaboratively building knowledge based on data.

Ivan describes the problem presentations as his favorite part of class because he can use his voice, compete, and exercise power:

> For today my favorite part was the homework. It looked pretty interesting and funny to me because Louis and our group would have convinced the whole class that our method was

right if you professor had not intervened. I like this part of the class maybe because there is room for discussion and also for competition. (Ivan, 2/27)

He describes his strategies when presenting problems:

I think that I'm really confident on my skills. I always make the others listen to me. So, confidence and determination … Also belief I think played a part. Believing in something that you think you are right, it gives you confidence and force to fight for it. (Ivan, 3/6)

Ivan invokes his middle-class white masculine subjectivity in his desire for competition, power and confidence. He likes that he was almost able to convince the whole class of his group's way of thinking and describes his discursive strategies to convince the class of his answers. He is confident and self-assured. He writes, "I always make others listen to me." For Ivan, physics is a prideful "fight" for "right answers", as opposed to a collaborative effort to deepen his and others' understanding of the physical world. He identifies with and perpetuates the elitism of physics as his positive physics identity.

Louis also engages in such discourse. However, his enthusiasm for physics stems from his experience in sports. Louis writes:

As an enthusiast of physics, I have been quite happy over these past two classes. I enjoy physics very much, because I find it to be a very practical science. That if you try to look physics is everywhere and it's fun. As a Mixed Martial Artist I notice the transfer of energy a lot. But I have a question. When I punch someone the energy transfers through their body. Though where does it go after that? Does it transfer through the floor or elsewhere? Just some things that's been on my mind. (Louis, 1/30)

Louis views physics as practical and sees physics in his life as a mixed martial artist. He says that "physics is everywhere." However, not all students feel this way. Martial arts is a male-dominated sport, and male-dominated activities are used often in physics instruction. For example, Louis wonders about the energy transfer in throwing a punch. Even though physics is relevant in activities deemed feminine, such as cheerleading, dance, and music, the conventional curriculum may not relate these activities to physics. Masculine topics in physics align with Louis's masculine interests, allowing him to take-up a positive physics identity.

Greg also sees physics in his gendered, raced, and classed activities, including downhill ski racing, rock climbing, and lacrosse. Wealthy white males are the primary participants in these sports. So, like Louis, these activities connect to his interest in physics. He writes in two examples the connection of physics to his extracurricular activities:

Physics is something in high school I had a blast in. Growing up as a downhill ski racer and a lacrosse goalie I learned what movements were needed to throw the ball to a teammate which I always felt helped me become better in sports. Especially with skiing and learning how to control my speed while going through gates to get the fastest time on the course. Having this information allowed me to make quick estimations on the sensations I would be feeling going through the race course. (Greg, 1/30)

A story of using physics I can think of is when I was rock climbing I was able to experiment with different grips by distributing my weight differently by moving my body in unusual positions. By doing this I'd be able to make more traction with the wall. And I found it really cool on how calculated the movements are even though no numbers were involved. (Greg, 2/6)

For Greg, there is a clear overlap between his pastimes and physics. Physics knowledge helped him as a lacrosse goalie in knowing where to throw the ball; as a skier in knowing how to control his speed; and as a rock climber knowing how to distribute his weight. Greg sees physics as related to these predominantly upper-class white masculine sports, allowing him to construct a positive physics identity.

Like Ivan and Louis, Greg's discourse of privilege, dominance, and entitlement reflect his middle-class white male subjectivity. He describes as a child being able "to take charge in a bit of a leadership position" which allowed him to become the goalie on his lacrosse team. He says he "plays to win" and "by being able to control and motivate [his] teammates by making big saves they would perform better for [him]" (Greg, 3/11). Greg's disposition suggests that he views himself as superior to his teammates. He engages with his teammates for his own benefit, so that "they would perform better" for him. His elitism, dominance, and self-centered goals reflect his subjectivity as a middle-class white male.

Greg's self-centered and dominant dispositions translate to performances in the physics classroom. Greg's description of working with two female classmates of color on a video analysis lab is similar to how he describes controlling his lacrosse teammates. He writes:

> I found I was able to make the program work the best and kept asking my team questions about what was going on to make sure that they understood what was happening as I explained the process of what I was doing. I wasn't sure if they understood or if they figured I knew what I was doing and just let me do my thing. (Greg, 2/13)

Greg dominates the group lab because he believes he makes the video analysis program work best. Though none of the students had prior experience with the software being used, Greg assumes his technological know-how is superior to his female peers of color. He positions himself as the source of knowledge and assumes they do not understand the assignment. This scenario is one illustration of how performances of subjectivities have implications for students' learning.

Ivan, Louis, and Greg demonstrate their middle-class white male subjectivities and positive physics identities. They make personal connections to the notion of physics from sports and other activities where they see strong physics connections. The three men take-up middle- class white male subjectivities in the physics classroom that also match their class status, race, and gender.

5.5.2 *Subjectivities Are Not Essential: A Woman Who Likes Physics and a Man Who Does Not*

Not all middle-class white males are dominant, confident, and self-interested, and not all females and minorities are collaborative, open minded, and selfless. Naira's performances do not match stereotypical notions of femininity. She takes control of her group, does not listen to her peers, and defends her thinking assertively. Unlike Ivan, Louis, and Greg, Naira's performances do not match her appearance. She is from Pakistan and has brown skin and long dark hair worn in a braid. Naira's performances show that females also engage in exclusionary practices, which may work to give them access to power that is traditionally reserved for men. Assimilating females into the culture of physics does not necessarily challenge the gender order but may engage them in an elitist discourse that sustains it. Naira, like Ivan, Louis, and Greg, enjoys physics. She writes:

> On Tuesday 1/28 I was afraid of physics, but today I feel as if physics can become one of my favorite subjects. I love how we can use technology to enhance our learning and with no doubt technology has succeeded in helping our understanding. Physics is fun, and I love how it involves physical techniques like using a meter rule, using a stopwatch, etc. I personally enjoy physics and wish to learn Physics102 as well, which includes electricity, power, gates, etc. What makes physics more fun is that it is very conceptual. (Naira, 1/30)

Naira is interested in topics associated with masculinity, including technology, physical techniques, and electricity. This contradicts an essentialized view of gender, where girls like animals and boys like machines, for example. Ivan enjoys the stereotypical cultural nature of physics, including competition, dominance, and elitism. Whereas, Naira enjoys the conceptual nature of physics, including using hands-on methods to study physical phenomenon. Naira's view of physics is more aligned with the practices of science (Etkina et al. 2006).

In addition, Naira sees physics as relevant to her life through nontraditional connections, including her lower-class work as a cashier. She writes:

> Doing physics can take any form when we walk, or work or even use our everyday routine we are "doing physics." An example can be found every minute of our life. Walking on a road as compared to walking on slippery ice is a form of physics. The concept involved in this example is friction. When we are at work, for example a cashier, he/she takes the money, puts it in and pushes the register to close it. In this example, pushing brings the concept of force while also involving power, energy and work as concepts. Using technology is a form of physics. Physics can be anything from switching on lights (concept of electricity) to even pulling a slinky (concepts of waves, motion, etc.). Physics is a part of our daily life and we don't even know it. (Naira, 2/6)

While Louis and Greg related physics to their pastimes, Naira sees physics in her workplace. Naira works part time as a cashier at a fast-food restaurant and describes the physics of operating the cashier machine. She describes physics as relevant to daily life that many do not notice.

Similar to Ivan, Naira's physics education began abroad and was defined by exams. She grew up in Pakistan. She writes:

> About my past experiences, I took physics, but it was much tougher. The physics course I took was called O'Levels from Cambridge University in England. The exam was 3 parts, it had 40 multiple choice questions, about 20 long and short questions, and an "Alternative to Practical" part, which was like 5 different labs done on the question paper and we had to analyze and answer questions. It was an extremely long exam, but I did well especially because I read the whole textbook and did 10 years of past exams as a practice. So, I think practice is highly important for any subject to prosper. (Naira, 4/10)

Naira discusses the rigor of her past physics course and the difficulty of the standardized exam. Such high stakes testing may promote fear, as Naira wrote on 1/28,

"I was afraid of physics." Naira attributes her success to practice, unlike Greg who assumes natural superiority and leadership in controlling his group's work. Successful male students may be viewed as having raw talent while successful female students are seen as quiet and hardworking (Carlone 2004). Students may embody this belief about themselves as well.

Naira, like Ivan, Louis, and Greg, performs characteristics associated with masculinity, including confidence and independence. In the following fieldnote excerpts, Naira prepares to present a homework problem with Sameer and two other female students. Their group was assigned part (A) of the following problem (Knight et al. 2009, p. 63):

> A light-rail train going from one station to the next on a straight section of track accelerates from rest at 1.1 m/s^2 for 20 s. It then proceeds at a constant speed for 1100 m before slowing down at 2.2 m/s^2 until it stops at the station. A) What is the distance between the stations? B) How much time does it take the train to go between the stations?

Naira stands at her seat leaning over the whiteboard, looking from her paper to the whiteboard as she copies down her work. Without looking up, she describes her approach out loud, "So we know velocity, the v is zero and we know vf … It's kind of complicated but I think it's correct." She writes and talks quickly without looking up. A female peer chimes in occasionally saying "yeah" to show her approval of Naira's work, but Sameer disagrees. He suggests, "But isn't it area?" Naira and Sameer exchange in a back and forth. Naira says, "But we don't know …" and Sameer says, "Isn't it?" Eventually, Naira shuts-down Sameer's suggestion and closes the discussion by saying, "We don't know. That's for this part. We already know the speed is the same." A female peer says, "Yeah." Naira continues writing. She continues to announce her thinking as she writes on the board: "And then 'a' is …" The second female peer remains quiet, watching as Naira writes. Sameer proposes a second time: "For some reason I get a different answer." Without changing her stance, leaning over the whiteboard with an Expo marker in hand, Naira looks up and affirms, "I actually got help from my sister." This interaction feels tense, and the first female peer tries to alleviate the discomfort of the confrontation by saying, "We will see when we go over it." A moment later, Sameer interrupts a third time: "But isn't it supposed to be the area under the curve?" The second female peer says, "I got confused" and the first female peer turns to explain to her. Meanwhile, Naira looks to Sameer and says: "You did it that way. Let me see." She looks over his work but continues to record her own ideas instead.

In this exchange, Naira takes charge of her group's presentation and attempts to work independently on their whiteboard. She announces her ideas as objective facts; her statements are not prefaced with "I think" or "I did it this way," nor does she ask her group members for suggestions or approval. Her dominant, confident, and assertive discourse is similar to that of Ivan, Louis, and Greg, and aligns with the notion of physics as authoritative and elite. Naira's confidence, assertion, and quick speed may suggest to Sameer and the two other female peers that physics is only understandable to a select few like Naira, and they are not included. Though Naira is female, her discourse of entitlement, confidence, and independence works to exclude others while promoting herself.

Sameer's discourse, on the other hand, is associated with traditional conceptions of femininity. Sameer is a male student from the Middle East. He interrupts Naira three times before she gives him her attention. He lightly nudges, "But isn't it area?" "For some reason I get a different answer." "But isn't it supposed to be the area under the curve?" He does not speak with the same conviction as Naira, and he is passive rather than assertive in his suggestions. He writes:

> Something that stuck out to me today is working with my group on the homework problem. I was listening to them say there are three parts to the problem, and in my mind I was saying excellent work. However, when it got to the point where we had to choose the formula, I stopped them and told them we cannot use the distance formula. The three insisted on using it anyway. I asked what answer they got and they told me 1730 m. Something is wrong... My group members said we are positive and I said I won't go against the three. I thought they could be right and I am wrong... After the professor stopped us when we were presenting, I knew I was right. But I learned one thing, that I must stick with my group members no matter because if I don't it will make me look like a stranger. It will even impact the teamwork and the enthusiasm of the group. (Sameer, 2/27)

Sameer mostly stayed silent while he observed Naira working on the whiteboard. According to his journal entry, he listened carefully to his group and kept his thoughts to himself, except when he thought they (i.e., Naira) made an error. Yet, his group mates' confidence (all three are females) and their majority opinion caused Sameer to doubt himself, when in fact they were wrong, and Sameer was correct. Sameer did not argue his way of thinking for what he considered to be for the benefit of the group dynamics. For Sameer, selfless collaboration was more important than the self-centered presentation of his thoughts. This viewpoint contrasts greatly with the words and actions of Naira, Ivan, Louis, and Greg, who each take control, argue, and believe they should be heard. His concern for others, complacency, and silence contrasts with a conventional culture of physics as independent, self-interested, and elite. In this moment, Naira's discourse associated with middle-class white masculinity dominated Sameer's modes of engagement, traditionally associated with femininity. In the end, Naira's ideas were presented on the whiteboard, and no one benefitted from Sameer's counter argument. Sameer writes about his decision to keep quiet:

> I agree that one must stand up for their point and disagree to a certain point; however, this never works in teamwork. When doing teamwork, there must be a consideration to everyone's answer and thoughts; however, at the end there must be an agreement, otherwise it won't be considered teamwork... I had this experience before in high school in physics class. I stood up in front of the class and said I got a different answer from my group members. Fortunately, I was correct, but I created hate from the group. They basically ignored me, but I apologized, and we became friends... Group work is sacrifice and agreement. [Even] if you are wrong, you at least created a team and an agreement and happiness to the group members. (Sameer, 3/6)

Sameer expresses his value of relationships and emotion and holds his independent thinking in order to create a team effort. In contrast to the views and performances of Ivan, Greg, Louis, and Naira, Sameer values the happiness of his group members over his pride in getting the right answer. Sameer's discourse and values do not align with his view of physics as authoritative and abusive. He writes:

> I HATE science so much! I really do. Back in my country I had the worst teacher ever. He used to beat up all the kids if they didn't know anything. He used a heavy stick made out of wood and beat students' hands. He hated me because he was conservative and since my family had American citizenship, he hated that fact. He used to call on me before he even teaches and asks me questions that I don't know. He made my life terrible. I hated science classes because of him. I had him as my science teacher for three years. (Sameer, 1/30)

Sameer's formative experience in science growing up in a Middle Eastern country involved judgment, discrimination, and abuse. He was expected to come to class already knowing, rather than come to class to learn. Corporeal punishment is unlawful in U.S. schools. Nonetheless, teachers may inflict harm by expecting students to know before even teaching them. Students with novice ideas may be viewed as lazy, stupid, or incompetent as opposed to ready to learn. Sameer's view of science as authoritative, unfair, and abusive does not align with his subjectivities as caring, collaborative, and selfless. Unlike Ivan, Greg, Louis and Naira, Sameer hates science, though his aversion may have nothing to do with the subject matter.

5.5.3 Subjectivities Are Dynamic and Gendered Performances Confer Status

Though Naira was domineering and assertive when working on her group's homework problem, she does not always perform roles associated with masculinity. During the last 4 weeks of the course, Greg and Naira were assigned to the same group with Sameer and another student. Greg described this group as "quiet," "intimidated," and "not as vocal," compared to his previous group with Ivan and Louis. Yet, these words do not describe Naira's performance when working on the homework problem presentation with her previous group. At times, Naira performs roles associated with masculinity, but other times she performs roles associated with femininity. The following fieldnote excerpt describes Naira's presentation of Part A of the light rail problem, and Ivan, Louis, and Greg's presentation of Part B. As a reminder, this is the problem:

> A light-rail train going from one station to the next on a straight section of track accelerates from rest at 1.1 m/s^2 for 20 s. It then proceeds at a constant speed for 1100 m before slowing down at 2.2 m/s^2 until it stops at the station. A) What is the distance between the stations? B) How much time does it take the train to go between the stations?

During problem presentations, the group presenting stands with their whiteboard in the front of the classroom and the rest of the class listens and can ask questions or make comments afterward. Sameer and his group members read the problem out loud, and Naira explains her thinking: "First I drew a graph … It accelerates at 2.2 and then slows down at 1.1." She explains that she worked on the problem in three pieces: speeding up, constant velocity, and slowing down. She found a distance for each segment and then added them to see how far the train went from one station to the next. She then states her final answer.

However, not everyone agrees. Louis, sitting in the back corner of the classroom, raises his hand and says, "I might nit-pick it because I did it my way. I used the equation $\Delta x = 1/2$ vf t." Naira calls back, "But we don't need the area," referring to the area under the velocity vs. time graph. A student in the middle of the room, calls out, "*B U R N* ...," drawn out and exaggerated, as if to say *she got you*! The class laughs. Ivan and Greg chime in. Naira responds, "No, I'm not finding the area under it." Tension builds in the class as students call back and forth.

Though her answer was incorrect, Naira's performance of confidence, independence, and self-assuredness may convince others that she is good at physics. Such discourse allows her to take-on Ivan, Louis, and Greg. During the semester, only one other female argues against these three male students. Had Naira shown humility and uncertainty, Ivan, Louis, and Greg may have taken the opportunity to express her ideas for her. Instead, she holds her ground to the surprise of the class (as indicated by a student's comment). Naira gains practice and skill in this discourse of independence and confidence. By the end of the semester, she writes, "I loved everything [in this course] but one thing would be homework presentation" (5/15). For Naira, homework presentations elevated her social status by giving her access to the power that Ivan, Louis, and Greg display. Her masculine performance allows her to defend herself from their critique, while positioning her as someone good at physics. This positioning had little to do with her understanding of physics, but depended on her discourse and performance as independent, assertive, and elite.

Naira does not enact the same competition, confidence and entitlement that Louis, Ivan, and Greg have when presenting their work on part (B) of the same problem. Greg explains their board and states their answer to the question. Naira disagrees with Greg, but only quietly mumbles: "I got 80 seconds." A couple other students mumble they got 80 seconds too, but no one calls out to "nitpick" their work. After a moment, a male student raises his hand and says, "You said you ..." He tries to repeat back how Greg solved the problem in an effort to find where he and Greg disagree. Naira's assertiveness followed by her silence suggests that she enacts masculinity to defend herself, but she does not use it as an offensive tool. Naira is outspoken when her own position is on the line. However, when Greg, Louis, and Ivan present their work, she keeps to herself. In fact, Naira is often quiet in problem presentations, except when she is presenting her own work. For Naira, the performance of masculinity is a tool she uses to her advantage at opportune moments, where patriarchal culture is rewarded with air-time, praise, and power.

5.6 Broaden Notions of Physics to Reach a Wider Range of Students

Three themes are presented in this study. First, middle-class white masculinity aligns with conventional notions of physics; second, subjectivities are not essential: a woman who likes physics and a man who does not; and third, subjectivities are dynamic and gendered performances confer status. These three themes offer several points for discussion about physics education and subjectivities. The findings of this study support Hughes' (2001) work in that students whose subjectivities align with the presentation of physics are more likely to take-up positive physics identities. A conventional notion of physics as elite, authoritative, and rational resonated with Ivan, Louis, and Greg. These elitist characteristics are stereotypical of middle-class white men. The alignment of white middle-class masculine subjectivities with conventional physics gave Ivan, Louis, and Greg the confidence to dominate whole-class discussions and thereby define physics in an interactive classroom.

However, not all middle-class white males engage in this way, and some females and minorities engage similarly and differently. For instance, Naira and Sameer show how students perform gender differently. Naira acts assertive, dominant, and self-interested when preparing her group's whiteboard for a problem presentation. Naira shares her interest in "physical techniques … electricity and power," which are stereotypically masculine science topics for study. Naira's narrative brings into question, must a woman "act like a man" to connect with conventional physics? In other words, does a student have to take-on interests, dispositions, and discourses associated with normative masculinity in order to be successful in conventional physics education? Because individuals take on femininities and masculinities independent from sex, some women delight in conventional physics and some men do not.

As a male physics student, Sameer did not embrace traditional masculinity. He considered the group dynamic over his individual achievement. His selflessness, insecurity, and care are associated with normative femininity. The mismatch of his identity and conventional physics may have contributed to his dislike of physics though he was talented at solving physics problems. Could physics have been constructed differently to be more collaborative, friendly, and kind? Would Sameer have connected with this notion of physics?

5.7 Implications and Conclusion

Even in an active classroom, physics may still be constructed as elite, authoritative, and simplistic. As students engage with each other, particular subjectivities are privileged in discussions and group activities. Three male participants and one female privilege competition and knowing how to do physics. In addition, the role of curriculum and instruction is paramount. Traditional textbook problems reinforce the notion of physics as abstract, irrelevant, and simplistic. In a student-centered and constructivist classroom, the instructor designs the intended curriculum and plays the role of facilitator or coach rather than source of knowledge (Driver et al. 1994). The ways the instructor's actions (or lack thereof) reinforce or push against conventional notions of physics is ideal for a future study.

In conventional physics education, a dichotomous notion of physics is created, where physics is elite (not accessible); independent (not collaborative); and competitive (not caring), to name a few. These terms are associated with middle-class white masculinity, while the words in parentheses are associated with femininity and are delegitimized in conventional physics education. This is problematic because it constructs a singular notion of what it means to do physics and who can do physics. This notion privileges middle-class white masculinity and leaves no room for students to negotiate their subjectivities to dis/connect with other notions of physics. It also discounts the importance of inclusion, collaboration, and care in the scientific endeavor. Therefore, physics teacher educators are tasked to broaden the ways that physics teachers think about physics and their students' multiple subjectivities. To support this desired outcome, teachers must be educated to critically examine constructions of physics and constructions of students' identities and to recognize the multiple and sometimes conflicting ways students negotiate their subjectivities in physics. Broadening and negotiating ways of doing physics that allows for multiple and differing subjectivities may encourage females and males to connect more in physics.

References

Brewe, E., Sawtelle, V., Kramer, L. H., Brien, G. E. O., Rodriguez, I., & Pamelá, P. (2010). Toward equity through participation in modeling instruction in introductory university physics. *Physical Review Special Topics – Physics Education Research, 6*(010106), 1–12.

Butler, J. (1988). Performative acts and gender constitution: An essay in phenomenology and feminist theory. *Theatre Journal, 40*(4), 519–531.

Carlone, H. B. (2004). The cultural production of science in reform-based physics: Girls' access, participation, and resistance. *Journal of Research in Science Teaching, 41*(4), 392–414.

Carlone, H. B., & Johnson, A. (2007). Understanding the science experiences of successful women of color: Science identity as an analytic lens. *Journal of Research in Science Teaching, 44*(8), 1187–1218.

Driver, R., Asoko, H., Leach, J., Mortimer, E., & Scott, P. (1994). Constructing scientific knowledge in the classroom. *Educational Researcher, 23*(7), 5–12.

Emerson, R. M., Fretz, R. I., & Shaw, L. L. (2011). *Writing ethnographic fieldnotes*. Chicago: University of Chicago Press.

Epstein, C. F. (1988). *Deceptive distinctions: Sex, gender, and the social order*. New Haven: Yale University Press.

Etkina, E., Van Heuvelen, A., White-Brahmia, S., Brookes, D. T., Gentile, M., Murthy, S., et al. (2006). Scientific abilities and their assessment. *Physical Review Special Topics-Physics Education Research, 2*(2), 020103.

Geertz, C. (2008). Thick description: Toward an interpretive theory of culture. In *The cultural geography reader* (pp. 41–51). London: Routledge.

Guba, E. G., & Lincoln, Y. S. (1989). *Fourth generation evaluation*. Newbury Park: Sage Publications.

Hestenes, D. (1987). Toward a modeling theory of physics instruction. *American Journal of Physics, 55*(5), 440–454.

Hughes, G. (2001). Exploring the availability of student scientist identities within curriculum discourse: An anti-essentialist approach to gender-inclusive science. *Gender and Education, 13*(3), 275–290.

Jones, M. G., Howe, A., & Rua, M. J. (2000). Gender differences in students' experiences, interests, and attitudes toward science and scientists. *Science Education, 84*(2), 180–192.

Karplus, R., & Brunschwig, F. (1969). *Introductory physics*. Captain's Engineering Services.
Kelly, A., & Sheppard, K. (2009). Secondary school physics availability in an urban setting: Issues related to academic achievement and course offerings. *American Journal of Physics, 77*(10), 902–906.
Knight, R., Jones, B., & Field, S. (2009). *College physics: A strategic approach* (2nd ed.). Boston: Addison-Wesley.
Kost-Smith, L. E., Pollock, S. J., & Finkelstein, N. D. (2010). Gender disparities in second semester college physics: The incremental effects of a "smog of bias". *Physical Review Special Topics - Physics Education Research, 6*(2), 020112.
McDermott, L. C., & the Physics Education Group, University of Washington. (1995). *Physics by inquiry*. New York: Wiley.
Merriam, S. B., & Tisdell, E. J. (2015). *Qualitative research: A guide to design and implementation*. San Francisco: John Wiley & Sons.
Pollock, S. J., Finkelstein, N. D., & Kost, L. E. (2007). Reducing the gender gap in the physics classroom: How sufficient is interactive engagement? *Physical Review Special Topics – Physics Education Research, 3*(010107), 1–4.
Rosa, K., & Mensah, F. M. (2016). Educational pathways of Black women physicists: Stories of life. *Physical Review Physics Education Research, 12*(1), 020113–020115.
Von Korff, J., Archibeque, B., Gomez, K. A., Heckendorf, T., McKagan, S. B., Sayre, E. C., Schenk, E. W., Shepherd, C., & Sorell, L. (2016). Secondary analysis of teaching methods in introductory physics: A 50k-student study. *American Journal of Physics, 84*(12), 969–974.
Wood, L. A., & Kroger, R. O. (2000). *Doing discourse analysis: Methods for studying action in talk and text*. Thousand Oaks: Sage Publications.

Chapter 6
Lecture Jokes: Mocking and Reproducing Celebrated Subject Positions in Physics

Anders Johansson and Maria Berge

6.1 Introduction

> If I could tag the air molecules – this is Joe, Joe, what are you doing? Generally, Joe is just going back and forth.

This lecture joke about air molecules comes from a list of quotes collected by students and presented on the website of Professor Ramamurti Shankar, who is a respected lecturer in physics at Yale University. One of the primary functions of a joke is to make people happy and invite laughter, but another effect is to create or support intimacy through acknowledging the shared knowledge of an in-group (Cohen 1999). At least some of the students clearly get Shankar's joke and are in this way invited in as physics insiders to some extent. But what are the conditions for this intimacy? This chapter takes a critical look at lecture jokes in physics, focusing on how they may reproduce norms that structure the possibilities of students' identifications. We analyse jokes collected by the first author during participant observation of lectures, together with the list of Shankar's jokes. In what ways do his and other lecturers' jokes reflect the culture of physics and shape identities within that culture?

We draw inspiration from studies of the culture and discourse of physics, which have indicated that physicists are often expected to exhibit traits that are generally perceived as male, such as competitiveness and authentic intelligence (Gonsalves et al. 2016; Hasse and Trentemøller 2008; Traweek 1988). Recent research has

A. Johansson (✉)
Communication and Learning in Science, Engineering Education Research,
Chalmers University of Technology, Göteborg, Sweden
e-mail: anders.l.johansson@chalmers.se

M. Berge
Department of Science and Mathematics Education, Umeå University, Umeå, Sweden
e-mail: maria.berge@umu.se

A. J. Gonsalves, A. T. Danielsson (eds.), *Physics Education and Gender*,
Cultural Studies of Science Education 19,
https://doi.org/10.1007/978-3-030-41933-2_6

pointed out how the discourse in physics classrooms may influence physics students, for example, by expressing limited ways of 'doing physics' (Danielsson 2009) in quantum mechanics courses (Johansson et al. 2018). Similar research in school science classrooms has shown that certain identities are attributed a 'celebrated' status, whereas others are regarded as secondary or even unintelligible (Archer et al. 2017a; Carlone et al. 2014). While some research has pointed to the role of jokes, humour and playfulness among physics students (Berge 2017; Hasse 2002, 2008), no attention has been paid to lecture jokes and how they, as part of the discourse in physics, may be part of shaping and reproducing physics culture. This chapter represents a first venture into analysing physics lecture jokes from an identity perspective. In particular, we aim to address questions about what values physics lecture jokes reflect, how this relates to gender and equity, and how it may structure students' possibilities for identification with physics and as physicists.

6.2 The Physics Community: Brimful of Humour

Part of talking scientifically is to be 'serious and dignified' all the time (Lemke 1990). In *Talking Science* (1990), Lemke elaborates on how scientific language has its own style and that 'scientific language' differs from 'everyday language'. The special style of scientific language is constituted by grammar, by word choice, by its use of idioms and metaphors, and 'largely by its avoidance of many stylistic devices that are freely used in other kinds of language' (p. 130). Nevertheless, the internet is full of humour related to science (Marsh 2016) and jokes are commonly used in science communication (Riesch 2015). Likewise, when students solve physics problems together, jokes account for a significant number of their interactions (Berge and Danielsson 2013; Due 2014; Scherr and Hammer 2009). Berge and Danielson (2013) also noticed that jokes among physics students can be an important way to demonstrate physics knowledge. In another study (Berge et al. 2012), which included both physics and bioengineering students, the physics students positioned themselves as insiders in relation to physics whereas the bioengineering students positioned themselves as outsiders, and this was partly done through jokes. The physics students made a joke about the formula $s = v{\cdot}t$ being an 'old classic', whereas the bioengineering students made ironic remarks about not being good enough, something that did not occur in the groups of physics students. A closer look at student laughter in physics problem solving (Berge 2017) revealed a more fine-grained pattern of physics humour. Humour and jokes contributed to a good working atmosphere (and thereby to the students' learning) but also interrelated with the disciplinary culture of physics. The students not only created and re-created humour that facilitated their social interactions, but through humour they also constructed local norms of science and engaged with the disciplinary discourse. In analysing physics problem solving among upper secondary school students, Due (2014) noticed a power dimension within the students' joking. Some jokes worked to exclude some students in the group, mainly by revealing their (lack of) physics knowledge. This is in line with Hasse's work (2002, 2008), which found several students excluded by jokes referring to science fiction and particularly to the number 42, a reference to *The Hitchhiker's Guide to the Galaxy* (Adams 1979). Barthelemy et al. (2016) reported sexism as well as sexist jokes in graduate physics and astronomy programmes. In this context, women sometimes experienced jokes as a form of microaggression from their classmates, for example: 'You like chickens for their legs. Not their brains' (p. 7). Here, humour became part of larger structures of gender discrimination. To sum up, wherever you find physics students, you will find humour and jokes of many different kinds.

According to Hasse (2002, 2008) there are several distinguishable elements within the community of physics, and one of these elements is a form of play. In her anthropological work in a first-year Danish physics programme, Hasse (2002) noted specific themes of interaction that were significant within physics education: namely the use of science fiction terminology, playing with physics equipment, and telling specific physics jokes. However, these themes were not distributed equally in the students' interactions. One group of male students used different forms of playing when interacting, while a large group of male students and most female students did not. Hasse concludes that gender cannot be the only differentiator here. In her ethnography of particle physicists, Traweek (1988) describes several unexpected skills that are crucial for being successful in the field. These skills are: being an informed gossip, managing to exchange judgements about one's peers, persuading colleagues to support one's work, managing news, and, in addition, 'being a competent performer of combative, tendentious jokes' (p. 121). According to Traweek, these jokes should preferably use technical language from particle physics to describe human behaviour; for example, a physicist jokingly told her that the physicists' exchange of information with each other was like 'photons being exchanged among interacting particles' (p. 121). Studying an academic subject does not only entail learning the subject matter, but also becoming part of a community and negotiating an identity (Johansson 2018a; Traweek 1988). To become recognised as a competent and legitimate participant, you need to be able to act like one. Thus part of becoming a physicist is to learn and deal with physics jokes.

6.3 The Complexity of Humour

Humour is a form of common knowledge in every culture: what counts as good humour differs from group to group and from moment to moment. That is why jokes are always conditional and why telling the right joke at the right time requires considerable cultural knowledge (Kuipers 2006). Humour can therefore be used to identify fellow members of a community through their appreciation (or not) of a joke, since a joke can never be explained without losing some of its entertainment value (this is one reason why several of the jokes presented in this chapter will appear dull after our examination). According to Cohen (1999) this is the foundation of the intimacy of jokes: it is the 'shared sense of those in a community' (p. 28). A joke is a specific form of humour, often described as something that consists of a setup and a punch line, where the punch line suddenly shifts the meaning in an unexpected way (Martin 2007). This form of joke is sometimes called a 'canned joke' since it is less conditional than other forms of humour and can work across different settings. But jokes and joking can also refer to informal jesting and witty quips. The quote from Shankar's homepage is an example of a mixture of a canned joke and informal humour. Jokes often deal with taboos and moral boundaries and at the same time mark social boundaries; to tag the air molecules and give them personal names is fun because it is irrelevant behaviour, and a form of play with a social boundary in physics education. Everyone who understands the joke is (more or less) pulled together automatically, but at the same time, those who do not understand it are shut out. Humour has many functions in social interaction. Kuipers (2006) points out that it can bring people together but also emphasise and augment differences in status; humour can shock, insult and hurt, and can consequently be used as an excuse for bad behaviour. This multiplicity of contradictory functions and meanings are at the core of humour – humour is and will always be contradictory. That is why humour also often touches upon social and moral boundaries; humour occurs in grey areas.

Although humour and jokes are often viewed as good, they must also be understood as based on and a part of the norms of any given social context, and they can thus be seen as having disciplining effects. Billig, in his critical evaluation of the role of humour in society criticizes both common-sense thinking and much academic research for sustaining an ideological positivism in relation to humour (Billig 2005). Rather, Billig argues, we must attend to the disciplinary role of humour, which through 'the possibility of ridicule ensures that members of society routinely comply with the customs and habits of their social milieu' (p. 2). Cohn's research (1987) on nuclear strategic thinking is an illustrative example of the disciplinary role of humour. She observed how the participants at a centre for defence technology and arms humorously talked about missiles without touching on the realities of nuclear holocaust that lay behind the words. For instance, the MX missile was called 'the Peacekeeper' and the bombs dropped over Hiroshima and Nagasaki were called 'Little Boy' and 'Fat Man'. This humoristic language limited what people working with nuclear weapons said and what it was possible to say, or even think, according to Cohn. In our treatment of lecture jokes, we aim to investigate the disciplining function of humour. With our poststructuralist understanding of social discourse, we also need to add that not only ridicule or demeaning jokes but also 'positive' or 'good-natured' jokes always represent a discursive structuring of what is funny, what is serious, what is possible to think and what is excluded from the social.

6.4 Jokes and Identity

In line with much research on gender and identity in science education, we use a feminist poststructuralist framework to conceptualize the social construction of identities. In particular we draw from the concept of *positioning* through discursive practices, as outlined by Davies and Harré (1990). Here, 'a discourse is to be understood as an institutionalised use of language and language-like sign systems' and 'discursive practices' refer to 'all the ways in which people actively produce social and psychological realities' (Davies and Harré 1990, p. 45). In this framework, identity or selfhood is achieved by taking up or constructing subject positions. In the words of Davies and Harré, positioning is 'the discursive process whereby selves are located in conversations as observably and subjectively coherent participants in jointly produced story lines' (p. 48).

We view jokes as one form of discursive practice contributing to the construction of subject positions. Every discursive practice draws from already established discourses. While this may seem to imply a deterministic structuring of available positions, there is significant agency possible through the ambiguity of speech acts. Whether actors take up or resist given positions is always a more or less open question. Jokes represent particularly ambiguous speech acts, which may simultaneously make fun of and accentuate given positions. Nevertheless, we argue that pervasive patterns in the form and content of jokes made in physics may serve to limit the available subject positions. One way of conceptualising this limiting is to point out how certain subject positions get framed as more 'celebrated' than others, and how taking up these positions gives access to status and power (Archer et al. 2017a; Carlone et al. 2014). In this study we follow the way Berge and Danielsson use positioning, which includes how physics content is positioned within the conversation. This is an unorthodox use of positioning, which usually has a clear focus on how individuals are positioned, that is, on subject positions. We argue, however, that it is appropriate to broaden the analytical focus to include the physics content since scientific language often avoids personifications, personalities and reference to individuals (Lemke 1990), and since the understanding of what physics is, is a dominant theme in physics conversations (Berge and Danielsson 2013).

Discourses may be institutionalised on many levels and around different topics (Davies and Harré 1990). For example, physics is commonly positioned as a difficult subject that requires a brilliant intellect, which is associated with masculinity (Francis et al. 2017). Persistent discourses about gender and physics serve to define the commonly imagined subject positions for physicists, and these may exclude many forms of femininity, meaning women have to perform as 'one of the boys' or craft other positions as competent physicists (Gonsalves et al. 2016). A feminist poststructuralist understanding of gender identity points to the discursive production of gender. A common way of conceptualizing this is to conceive of gender as produced through performative acts, discursive practices where subject positions are mostly reproduced in line with given patterns, but where a possibility of change lies in 'subversive' or 'wrong' repetitions of expected gender behaviour (Butler 1990). We consider jokes as performative speech acts that are part of constructing gendered subject positions. In our analysis, we focus on the construction of subject positions in lecture jokes. In particular we ask how physicists *and* physics are positioned in the jokes.

6.5 Getting the Jokes: Materials and Methods

The material for our analysis consists of two datasets. First, we use a list of jokes collected by students in the introductory physics classes taught by R. Shankar at Yale University, and then published on his website (Shankar 2015). Shankar has long taught physics, and we understand the long list of quotes to be a sign of admiration from his students. Second, we use field notes of utterances interpreted as jokes in observations of physics classes conducted by the first author[1] (see Table 6.1). The extracts used for this study are quotes from teachers noted more or less verbatim in the notes taken during observations.

In the case of Shankar's quotes, the students defined (and edited) what he said as humour. In our data based on observations, the first author picked out humorous sequences that caught his attention. In most cases this was determined by laughter from the students or the lecturer, a clear indication of humour being performed (Berge 2017). Most of the humour analysed here is therefore co-produced (Söderlund 2016) with the students in some way. In the case of Shankar, co-production occurs through the students' role as editors; in the fieldwork setting, by students laughing along with the lecturer. While aware that the quotes from Shankar and university lecturers are a mix of canned jokes and informal humour, we will, for convenience, label all analysed extracts 'jokes'. We are also aware of the fact that 'the same sentence can be used to perform several different speech acts' (Davies and Harré 1990, p. 50), which makes our analysis of Shankar's (edited) jokes more difficult since we do not have access to the whole context (much more context is available for the observations). The ambiguity of jokes makes analysis even more complicated. However, according to Davies and Harré, nobody can have full access to a conversation and all conversations can be perceived in several ways. What we have attempted to capture in our interpretation of the material is thus dominant constructions of subject positions as physicists, recognising that other interpretations are possible and that these positionings could be taken up in different ways in the participants' negotiations.

Table 6.1 The number of recorded jokes, courses, classes and teachers in all of the data sets

Material	Jokes	Courses	Classes	Teachers
List of quotes from R. Shankar	111	Unknown	Unknown	1
Observations of quantum physics classes at two Swedish universities	48	3	29	5
Observations of various classes at two Master's programmes in physics at two Scandinavian universities	13	5	7	5

[1] Johansson did ethnographic field work in physics classes to answer questions about identity and culture in physics education. Those projects also involved interviews with students, but the material analysed here is the notes from the observed physics classes. For further discussion of the methodology, see Johansson et al. (2018) and Johansson (2018b, c).

We employed an open coding procedure to analyse the jokes, with both authors initially coding parts of the material. In this process it was helpful that both authors have a background in physics. On the basis of this analysis, we developed a selection of codes characterizing how physics and physicists were positioned within the jokes. The refined coding scheme involved codes such as 'physics is rigorous/breaking the rules' and 'physics does not concern ordinary things'. Often a joke positioned both physics and physicists in some way. For example, a lecturer presenting the postulates of quantum mechanics said, 'It is a postulate – we don't know where it came from. And we don't care'. We interpreted this as positioning physics as abstract and concerned with very specific things, thus implying that physicists are 'narrow-minded' in some sense. After developing our coding scheme, we independently coded all the jokes according to it and discussed both our diverging and converging interpretations of the gist of the jokes and the positioning done in them. This second iteration led to our expanding the scheme as we could see that a few more positionings were being done. In this way we arrived at a consensus interpretation for most of the jokes and could summarise our interpretations in the three major themes presented in the following sections: Physics is serious and rule-bound; Physics is difficult and physicists are smart; Physics is interesting and physicists are nerdy and passionate.

During our analysis we found that the jokes differed in a few ways. For example, Shankar's jokes concerned college physics, whereas the university lecturers' jokes concerned upper-division university physics. However, we also noticed local details within the jokes, like references to being a Harvard student (Shankar is at Yale) or to a specific door that was difficult to close (Master's course). We also noticed that some jokes reappeared in quantum physics courses, almost like canned jokes. For instance when half of the students answered 'yes' and the other half 'no', two lecturers joked about making 'statistical averages', similar to quantum mechanics calculations. A category of jokes that occurred in both datasets concern the teaching situation, but we have chosen to not focus on these here. Shankar made several jokes about himself, positioning himself as the Einstein stereotype, both absent-minded and wise (not answering email but understanding the meaning of life). The lack of mean jokes was striking; we found no jokes directed towards any student to diminish her or him, as occurred in student-student interactions in Due's study (2014). Personal nouns were almost non-existent. In fact, in the list of Shankar jokes, a physicist is explicitly gendered only once: 'Let's say the physicist gets stuck while climbing, and you want to send him something. It may be food, or since it's a physicist, he might say "Send me my Wolfson and Pasachoff (our textbook)! I haven't read it in two days!"' Still, we did find consistent patterns of positionings of physicists, and it is to our presentations of these that we turn in the following three sections.

6.6 'Never trust a log plot': Physics Jokes Are Serious Business

The majority of the jokes concerned the physics content itself, or the process of learning and doing physics. Geelan (2013) noticed that many high school physics teachers use 'dry humour' in year 11, and we could see a similar pattern here. Examples of dry humour are Shankar's description of the risk of trusting a log plot: 'Never trust a log plot. And especially never trust a log log plot' and a quantum mechanics lecturer's comment, 'I'm gonna do what I always like to do, give you a new operator.' We categorise these jokes as dry humour because although the content is serious and positions physics as closely related to mathematics, there is a subtle comedic delivery. Another form of humour in talking about physics content is the use of absurd scenarios and strange metaphors (Berge 2017). Shankar illustrated the Einstein equivalence principle by suggesting the idea of switching Earth for another planet:

> Say you're in an elevator. I could do two things to you and you wouldn't know the difference. I could pull the elevator up with a rope and you'd begin to feel heavy. Or, I could replace the planet beneath you with a bigger planet and you'd feel heavy. Now most likely I'll do the first one. But you can't tell the difference!

Or similarly, on the idea of eternal life, 'If you live 15 billion years, then you will be able to see the back of your head'. Common to both these examples is the fact that what is actually possible is not relevant, but the ideas are. Shankar is playing with variables while illuminating the physics he is lecturing about.

Physics at this level is often abstract, with few realistic contextualisations. Shankar, however, is able to joke about that too: 'Many people think that, since they're going to be doctors or something, they're never going to need to know about relativity. Well, what if one of your patients starts running away from you at the speed of light? Then you really need to know this.' Here Shankar manages to make light of the fact that the concept of speed of light is not applicable in everyday life while giving the students an absurd (and entertaining) picture of imaginary patients taking off at the speed of light. These absurd scenarios or strange metaphors are used as tools for contextualising physics and promoting learning.

The teachers also joked by breaking informal norms about scientific language, in the same way as the students in Berge's study (2017). As in our first example, the air molecule called 'Joe', this kind of humour is based on the informal norm that certain things (like personal names) are of no significance in physics discourse. Rather than playing with physics concepts, the teachers here play with physics norms, and often with what Lemke (1990) refers to as the stylistic norms of science. Lemke has summarised the common style of talking science into a list of nine norms that he describes as a 'recipe for a dull, alienating language' (p. 134). One of the norms on the list is to be 'serious and dignified' when talking science, something Shankar does not seem to care too much about judging from the analysed quotes. Shankar also violates other norms on Lemke's list, like the norm of avoiding the use of personifications and dramatic accounts. This is clear when he talks about the mass of the Earth: 'The Earth's whole mass – you, me, China – everything is pulling it down', and also when he describes the trajectory of a rocket:

> Say you're firing a rocket launcher. What angle should you fire it at for maximum range? Say you fire it straight up. The good news is that it's going to be up in the air for a very long time. The bad news is that it's going to land on your head.

Similarly, one quantum mechanics lecturer talked about the 'violent name' of an operator (the annihilation operator), and the students laughed. This kind of humour intrudes upon the norm of avoiding metaphorical and figurative language when talking science, and especially avoiding emotional, colourful, or value-laden words (Lemke 1990, p. 133). Likewise, when a lecturer says that 'But that's when we talk about space-like, separation, distance. Now it's time-like [...] When this gets bigger, this gets smaller, that is the hand-waving explanation' this intrudes on the norm of being as verbally explicit and universal as possible. By making jokes like this, these norms become strengthened; the laughter reinforces the point that we are not, in this context, supposed to talk this way. Thus, humour is a way to make a norm or an implicit rule explicit in the conversation. Likewise, this kind of humour can be a way to teach the students awareness of these norms, since humour is one way to make a norm or an implicit rule explicit (Berge 2017).

We know from previous research that students pick up on these examples of physics humour. When they solve physics problems together, they use absurd scenarios both in order to be funny and to clarify what they mean in order to be able to solve the task (Berge 2017). Making an accurate joke about physics content can even be a way for students to display their knowledge to their teachers: being able to make the 'right joke' implies that you have the 'right physics knowledge' (Berge and Danielsson 2013). Within the jokes described above, physics is positioned as a subject that has certain rules and norms and a special style that is not to be neglected; everyday words like 'violent' and 'China' become funny in this context. The physicist, on the other hand, is positioned as someone who appreciates this kind of somewhat dry humour. This may work to exclude students who are not necessarily used to joking in this way (Hasse 2002).

6.7 'It's Not Fun, It's Not Easy, But It's in the Notes': Hard Physics, Smart Physicists

Another theme in many of the analysed jokes was that physics is hard and difficult. This was sometimes explicitly stated, as by Shankar when he talked about a new topic in physics: 'This is very different from a graduate quantum course which I could teach in my sleep and which you could listen to in your sleep. Here, everyone needs to be awake – this causes some added difficulty.' Here Shankar specifically mentions that even he needs to concentrate, implying that this is not easy for anyone. Similarly, the university lecturers made jokes about how difficult physics can be: 'You stick it in here, you do the integral; it's not fun, it's not easy, but it's in the notes'. On one occasion, a lecturer presented a homework task, saying, 'It's more difficult than the last one.' This statement was met with laughter from the students and the lecturer continued, 'I realize it's more than a little harder,' making the first description an understatement. This is the core of the fun in many of these jokes: physics is so difficult that calling it easy must clearly be a joke. Statements like 'this feels easier'[2] and 'this is a standard problem'[3] are followed by laughter. A quantum physics lecturer joked on the same theme when talking about Einstein's derivation of the photoelectric effect: 'So, this was pretty easy. For this simple derivation he received the Nobel Prize in 1921 … No, there were some other things as well.'

Physics, like mathematics, has a certain status because it is seen as difficult (Archer et. al 2017b; Francis et al. 2017; Gonsalves and Seiler 2012). Anyone who studies these subjects is assumed to gain some of this status, as one of Shankar's jokes illustrates:

> Mathematicians are always ahead of physicists, and physicists are always a little bit ahead of engineers, although that difference is not always clear anymore. It's because it takes so much time for our president to catch up with everything. He says 'How many barrels of oil will we save by you studying quantum mechanics?' and then we say 'Well, zero barrels' and he gets confused. So either you find this quantum stuff very useful or just use it to scare the hell out of everyone else.

In this joke, mathematicians have higher status than physicists and engineers (the opposite is illustrated in other jokes), but the main message is that studying quantum physics is a way to improve your position in society, even in relation to the president. Studying physics means you are already doing something difficult so that, ironically, it is often less important for physics students to gain the highest grades in examinations at university level (compared to other prestigious programmes like law and medicine). For engineering physics students, failing examinations is not considered extraordinary; simply passing can be considered success (Nyström et al. 2019). This acceptance of difficulty can be recognized in the American college context as well. Shankar guessed that some of his students had not got everything in his physics lesson and told them that he accepted this: 'It's okay if you don't get it, because if you all do get it, then I'm out of a job. I rely on you guys not getting it.' On another occasion Shankar acknowledged (and legitimised) that some students could not follow everything in class 'We're going to go over this again, as part of our No Child Left Behind program. Some children were left behind Wednesday; I know, because I saw lots of puzzled faces.' The point is that as long as you learn this difficult subject, you will at least be better than other people, who have not studied physics at all: 'There is one congressman who knows physics, and he's just bullying everybody around, because when he writes an equation down, none of the rest of them know what to do!' By positioning physics as difficult you simultaneously

---

[2] After writing the simplified Schrödinger equation with the given substitutions, the lecturer says 'ok … this feels easier, doesn't it?' and laughs a little. The equation is still fairly long. None of the students respond.

[3] The lecturer says that the third homework problem is available and that 'this is a standard problem' (compared to the earlier very tricky one). Several students laugh.

position physicists as smart. In fact, in these discursive positionings, anyone who manages to study physics is smarter than anyone else.

Even though students can be positioned as smarter than the rest of the world by studying physics, the university lecturers struggled to get them to ask questions. When one of the lecturers demonstrated an example on the blackboard and called for questions, the lecturer couldn't resist joking about the silence from the class: 'I'm gonna take that as meaning I was extremely clear and you understood everything.' This statement was met with laughter because the silence obviously did not indicate understanding. This lecturer also joked about the students' silence, saying, 'the floodgates are open so …' implying that the students seldom asked questions, and also that 'I will not be mean if you answer incorrectly.' Clearly the students were afraid to ask stupid questions.[4] Maybe being implicitly positioned as clever (just by being present in a physics classroom) makes it even harder to risk being perceived as stupid, since the fall is higher. This does affect the possibilities for learning within these classrooms at university level, since exposed confusion can be a valuable resource for learning physics (Dowd et al. 2015).

6.8 'It's Fun, and If You're Lucky, You Might Actually Get Paid!': Interest, Nerdiness, and Devotion

Physics is not only positioned as hard and physicists as smart, but the jokes also often played with notions of the engaging nature of physics and the engaged, or even nerdy, interest of physicists. In our analysis, we found that physics was positioned as a subject for interested and engaged students. Every teacher naturally wants to make their subject as engaging and interesting as possible for students. In physics, this can be done by engaging demonstrations. For example, a graduate level quantum mechanics lecturer used balloons to illustrate wave function collapse. After 'collapsing' a balloon, the lecturer commented that 'this wasn't meant to be a real analogy, though; this was just for fun', which generated some laughs from students, supposedly in part about the physics idea of 'fun'. But the serious business of physics cannot necessarily be learned through such 'engaging' demonstrations, and physics teachers may have to convey what fun physics is in other ways. Shankar told the students outright: 'That's why I'm telling you all to go do physics for the rest of your lives. It's fun, and if you're lucky, you might actually get paid!' This joke reflects the possibly fierce competition for jobs in the academy, but in particular it positions physicists as dedicated people. They may do physics their whole lives because it is fun; it is more important than money.

[4] This topic was brought up in several of the interviews that were part of the projects conducted by the first author. The interviewed students said that they often found avoiding asking stupid questions to be important in the physics classroom.

Other jokes positioned physics as the only really important thing in the world. A typical joke from Shankar exemplifies this: 'This is a very important day. You can forget your birthday, forget anniversaries, but you need to remember this day, because this is the day that you will learn Newton's Laws.' While this is an explicit instruction about how students should relate to physics, Shankar also conveys this message by telling (hyperbolic) anecdotes from his own life: 'I forgot what my life was like before quantum mechanics. I know I was playing in a sandbox and someone was trying to beat me up, but I don't remember when that was.' This same pattern can be seen in jokes about 'physicists in general'. For example, Shankar joked about physicists' hobbies: 'This problem in your book says that a physicist is hiking up the Alps. You know that's a joke, right?' The joke here lies in pointing out the irony of a hiking physicist, positioning interests other than physics as unthinkable for physicists.[5] Some jokes indicate that some areas of physics are seen as particularly interesting or sexy, which is not necessarily mirrored in students' experiences of learning them (Johansson 2018a; Johansson et al. 2018). For example, in one of the quantum mechanics courses the students were asked if they felt 'psyched up' for the subject in the first lecture, to which several of the students responded positively (see Johansson et al. 2018 for further discussion of this). In pointing out how interesting his current class is, Shankar says: 'Relativity and quantum didn't used to be taught in this class, which is a shame, because they are two of the sexiest topics in all of physics.'

When the only interest of physicists is physics, that is also where they find their joy. If physics is simultaneously perceived as dry and difficult, such enjoyment positions the physicists as nerdy in some way for finding pleasure in narrowly focused interests. An example of this is how Shankar describes his own pleasure in drawing ray diagrams:

> When I was a student I used to just draw two rays and be done with it, but now that I am nearing retirement I am so excited to draw all these different rays and see that they all hit the same spot. You guys don't know how much pleasure this gives me.

Similar positioning was done, albeit in a more low-key way, by a tutorial teacher in quantum mechanics. After having spent some time on the derivation of a complicated expression that many students may have had trouble following, this teacher says that it would be easy to come up with a more specific solution for the problem at hand but that, 'for fun, I have showed you the general expression.'

These examples indicate how physicists and physics students are expected to have a very deep engagement and enjoyment of physics. At the same time, the narrowly defined physics is positioned as very important, nothing else matters. This is evident in how Shankar describes 'the beauty of physics':

> In this first problem, there is a car driving along a cliff, and the car just jumps off. This person has decided to end it all. Now, we want to know at what time the car hits the ground.

[5] This does not match our personal experiences of cultural attitudes to outdoor life among physicists in Sweden. It is not that uncommon for physicists to be outdoor enthusiasts.

> This is the beauty of physics, because if this were a psychology class we'd want to know why the person was jumping, but we are simply concerned with how long it takes.

In positioning physics and physicists as unconcerned with complicated things like human emotions, Shankar reflects the spirit that Traweek describes as promoted by physics textbooks: Care about the fundamental things and ignore things like '"cute" and/or irritating creatures, from nude females to fleas' (Traweek 1988, p. 80). Several of the lecturers also emphasized that the more pure, fundamental, or theoretical a statement is, the more beautiful it is. Shankar: 'I just love this problem, because it has no numbers! I mean, here's mu, here's B, and everybody's happy!' A quantum mechanics lecturer: 'Theoretical physicists have a solution for all these $\hbar$. They put them to one [laugh from students]. But I try to keep them around in class.'

When physics is constructed as concerned with only the most fundamental things, this gives physicists licence to be bad at the things ordinary people manage, such as social interactions (see Willey and Subramaniam 2017). One example is drawing 'nice' pictures. Shankar joked about his abilities after drawing a stick figure of a 'ballerina': 'I guess it's better to try and fail than not to try at all.' Similar positioning was done by a quantum mechanics lecturer, who after failing to draw a circle on the blackboard using a compass, drew one by hand and said: 'Well, technology … it doesn't matter that much how it looks, really.' Students also participate in these discourses. In one of the quantum mechanics classes the lecturer asked a question, but halted in mid-sentence to reflect upon the phrasing, which sounded weird. The lecturer then asked, 'Is this how you say it?' A short discussion among the students was ended by one of the students saying, 'We are physicists, don't bother about that; you can go and study languages or something in that case,' whereupon many students laughed.

6.9 The Punch Line: Discussion and Conclusions

In our analysis of lecture jokes, we have seen how conventional discourses about physics and physicists are largely reproduced, at the same time as they are joked about. The expressed jokes display a language full of absurd analogies, abstractions and (sometimes intentionally bent) rules on how to speak. The language is often straightforward; we found no euphemisms, and in contrast to the studies of Cohn (1987) and Barthelemy et al. (2016), gender was almost never made explicit in our data or even alluded to. The relative 'formality' or 'dryness' of the joking in physics lectures points to how physics discourse and physics culture is conceived of as value-free, supporting the notion of physics as a 'culture of no culture' (Traweek 1988, p. 162). Nevertheless, as Traweek and others have emphasized, this culture *does* have values and expectations that may be excluding many people. In the context of physics lectures, we claim that the discursive positionings of physics and physicists we have outlined in this chapter are part of structuring the possibilities for students' identifications in physics, even though the signals are not as strong as might be the case in more sexist discourses (Barthelemy et al. 2016; Cohn 1987).

Another result of our analysis, which is in line with previous research, is that physics is positioned as difficult and the physicist as smart (Francis et al. 2017; Gonsalves and Seiler 2012; Traweek 1988) with a narrowly-focused passion for physics and nothing else. The physicist is also positioned as someone who appreciates a specific kind of humour (which we know can be a gendered position, see Hasse 2002). Again, the jokes in this study are both including and excluding: at the same time as the jokes legitimise a passion for physics, they implicitly exclude students with a more moderate love for the subject. Likewise, although everyone in the room may feel part of the physics community, insider jokes differentiate the students from those who do not understand them. We note a risk here that physics jokes may, like science comedy, fall into ridiculing people who do not like or understand science 'playing on the superiority aspect of humour' (Riesch 2015, p. 773). The celebrated subject position is in other words narrow, sharing many similarities with the position of the nerd or geek (Johansson 2018c; Willey and Subramaniam 2017).

The material analysed in this chapter does represent jokes in the rather formal context of lectures, where civility is presupposed. Therefore, our material represents not only a limited number of physics jokes, but a very specific sample of them. However, although this study represents a first attempt to investigate the role of lecture jokes in physics education, it has been striking to note the commonalities of physics joking across several courses, lecturers, universities and countries. This is something that we take as pointing to a shared physics culture. We do believe that the positionings enacted through the discourse of jokes contribute to students' identifications as physicists. One piece of evidence for the effect of these discourses lies in the uptake of these forms of joking and subject positions among students. For instance, we have acknowledgement of the rules of science by joking dryly around them (Berge 2017), but also the discourse of 'physicists don't need to care about other stuff'. Continued research could shed more light on the role of jokes in teaching and learning physics, and how this aspect of the disciplinary culture restricts or opens possibilities for various positionings. Interesting lines of inquiry would be to look in more detail at how jokes are both delivered and received, as has been done with conversation analysis of video-recorded science lessons by Roth et al. (2011), or to compare joking in different disciplines or stages of education.

Lecture jokes have both benefits and drawbacks. We know that teachers' use of humour and laughter is associated with learning (Banas et al. 2011; Roth et al. 2011). Nevertheless, our study also illustrates a pattern where the teacher may, with the very best intentions, be excluding some students without being aware of it. Shankar's joke about physicists who never hike can be interpreted as an example of that. Another drawback of the jokes analysed is that jokes that position students as smart may increase the risk of students feeling like frauds, the 'impostor syndrome' which has been shown to influence women in physics (and other subjects) more than men (Ivie et al. 2016). The jokes based on the lack of students' questions in our data are evidence of these feelings in the classrooms. However, in our ethnographic data we saw how humour could also have a positive effect in a classroom and change the atmosphere for the better. For example, when a lecturer made a joke by confessing that a physics problem needed extra attention and that even the lecturer would need to look stuff up, this was followed by students' laughter and less silence in the classroom. In another situation, the same lecturer, while waiting for answers from the students, explained that the question was difficult. A student responded to this by bursting out 'this [the physics] is starting to get weird' and everyone laughed again. This time, a student's humour became an icebreaker that opened up the interaction in the classroom. Both these instances are examples of the valuable role that humour can have in increasing closeness and involvement in classrooms (Banas et al. 2011).

Humour is an ambiguous form of communication, which can have both positive and negative effects on the classroom, and is also an efficient way to make norms or implicit rules explicit in the conversation. Thus we cannot make general recommendations about the use of joking and humour in physics lectures. However, as teachers we should reflect on the power of humour when we use jokes in the classroom, and on what discursive positionings the jokes we tell enact. In that way we can take small steps to remove potential barriers to making physics open to all.

References

Adams, D. (1979). *The Hitchhiker's guide to the galaxy*. London: Pan Books.

Archer, L., Dawson, E., DeWitt, J., Godec, S., King, H., Mau, A., et al. (2017a). Killing curiosity? An analysis of celebrated identity performances among teachers and students in nine London secondary science classrooms. *Science Education, 101*(5), 741–764. doi: 10/gctkbf.

Archer, L., Moote, J., Francis, B., DeWitt, J., & Yeomans, L. (2017b). The "exceptional" physics girl: A sociological analysis of multimethod data from young women aged 10–16 to explore gendered patterns of post-16 participation. *American Educational Research Journal*, 54(1), 88–126. doi: 10/gctkdj

Banas, J. A., Dunbar, N., Rodriguez, D., & Liu, S.-J. (2011). A review of humor in educational settings: Four decades of research. *Communication Education, 60*(1), 115–144. doi: 10/fr95jf.

Barthelemy, R. S., McCormick, M., & Henderson, C. (2016). Gender discrimination in physics and astronomy: Graduate student experiences of sexism and gender microaggressions. *Physical Review Physics Education Research, 12*(2), 020119. doi: 10/gctkc7.

Berge, M. (2017). The role of humor in learning physics: A study of undergraduate students. *Research in Science Education, 47*(2), 427–450. doi: 10/f3pdvv.

Berge, M., & Danielsson, A. T. (2013). Characterising learning interactions: A study of university students solving physics problems in groups. *Research in Science Education, 43*(3), 1177–1196. doi: 10/f2z3vm.

Berge, M., Danielsson, A. T., & Ingerman, Å. (2012). Different stories of group work: Exploring problem solving in engineering education. *Nordic Studies in Science Education, 8*(1), 3–16. doi: 10/gd4s2p.

Billig, M. (2005). *Laughter and ridicule: Towards a social critique of humor*. London/Thousand Oaks: Sage.

Butler, J. (1990). *Gender trouble: Feminism and the subversion of identity*. New York: Routledge.

Carlone, H. B., Scott, C. M., & Lowder, C. (2014). Becoming (less) scientific: A longitudinal study of students' identity work from elementary to middle school science. *Journal of Research in Science Teaching, 51*(7), 836–869. doi: 10/f6dgcr.

Cohen, T. (1999). *Jokes: Philosophical thoughts on joking matters*. Chicago: University of Chicago Press.
Cohn, C. (1987). Sex and death in the rational world of defense intellectuals. *Signs, 12*, 687–718. doi: 10/cvbz3v
Danielsson, A. T. (2009). *Doing physics – doing gender: An exploration of physics students' identity constitution in the context of laboratory work*. PhD thesis. Uppsala University. http://urn.kb.se/resolve?urn=urn:nbn:se:uu:diva-98907.
Davies, B., & Harré, R. (1990). Positioning: The discursive production of selves. *Journal for the Theory of Social Behaviour, 20*(1), 43–63. doi: 10/fxdfdt.
Dowd, J. E., Araujo, I., & Mazur, E. (2015). Making sense of confusion: Relating performance, confidence, and self-efficacy to expressions of confusion in an introductory physics class. *Physical Review Special Topics – Physics Education Research, 11*(1), 010107. doi: 10/gc5s8c.
Due, K. (2014). Who is the competent physics student? A study of students' positions and social interaction in small-group discussions. *Cultural Studies of Science Education, 9*(2), 441–459. doi: 10/f25bmc.
Francis, B., Archer, L., Moote, J., DeWitt, J., MacLeod, E., & Yeomans, L. (2017). The construction of physics as a quintessentially masculine subject: Young people's perceptions of gender issues in access to physics. *Sex Roles, 76*(3–4), 156–174. doi: 10/f9nnjt.
Geelan, D. (2013). Teacher explanation of physics concepts: A video study. *Research in Science Education, 43*(5), 1751–1762. doi: 10/gd4tnh.
Gonsalves, A. J., & Seiler, G. (2012). Recognizing "smart super-physicists": Gendering competence in doctoral physics. In M. Varelas (Ed.), *Identity construction and science education research* (pp. 157–172). Rotterdam: Sense Publishers. doi: 10/cs4t.
Gonsalves, A. J., Danielsson, A., & Pettersson, H. (2016). Masculinities and experimental practices in physics: The view from three case studies. *Physical Review Physics Education Research, 12*(2), 020120. doi: 10/f3rnhx.
Hasse, C. (2002). Gender diversity in play with physics: The problem of premises for participation in activities. *Mind, Culture, and Activity, 9*(4), 250–269. doi: 10/dsr9f3.
Hasse, C. (2008). Learning and transition in a culture of playful physicists. *European Journal of Psychology of Education, 23*(2), 149–164. doi: 10/cxg4zs.
Hasse, C., & Trentemøller, S. (2008). *Break the pattern!: A critical enquiry into three scientific workplace cultures: Hercules, Caretakers and Worker Bees*. Tartu: Tartu University Press. https://cordis.europa.eu/publication/rcn/10449_en.html.
Ivie, R., White, S., & Chu, R. Y. (2016). Women's and men's career choices in astronomy and astrophysics. *Physical Review Physics Education Research, 12*(2), 020109. doi: 10/gc5vqt.
Johansson, A. (2018a). Undergraduate quantum mechanics: Lost opportunities for engaging motivated students? *European Journal of Physics, 39*(2), 025705. doi: 10/gctj9n.
Johansson, A. (2018b). *The formation of successful physics students: Discourse and identity perspectives on university physics*. PhD thesis. Uppsala University. http://urn.kb.se/resolve?urn=urn:nbn:se:uu:diva-357341
Johansson, A. (2018c). Negotiating intelligence, nerdiness, and status in physics master's studies. *Research in Science Education*. doi: 10/gfjxxr
Johansson, A., Andersson, S., Salminen-Karlsson, M., & Elmgren, M. (2018). "Shut up and calculate": The available discursive positions in quantum physics courses. *Cultural Studies of Science Education, 13*(1), 205–226. doi: 10/f3rq2b.
Kuipers, G. (2006). *Good humor, bad taste: A sociology of the joke*. Berlin: Mouton de Gruyter.
Lemke, J. L. (1990). *Talking science: Language, learning, and values*. Norwood: Praeger.
Marsh, O. (2016). 'People seem to really enjoy the mix of humour and intelligence': Science humour in online settings. *Journal of Science Communication, 15*(02). doi: 10/gd22z4.
Martin, R. A. (2007). *The psychology of humor: An integrative approach*. Burlington: Elsevier Academic Press.
Nyström, A.-S., Jackson, C., & Karlsson, M. S. (2019). What counts as success? Constructions of achievement in prestigious higher education programmes. *Research Papers in Education*, 34(4), 465–482. doi: 10/gc8vmx.

Riesch, H. (2015). Why did the proton cross the road? Humour and science communication. *Public Understanding of Science, 24*(7), 768–775. doi: 10/gd23sr.

Roth, W.-M., Ritchie, S. M., Hudson, P., & Mergard, V. (2011). A study of laughter in science lessons. *Journal of Research in Science Teaching, 48*(5), 437–458. doi: 10/fh7z7v.

Scherr, R. E., & Hammer, D. (2009). Student behavior and epistemological framing: Examples from collaborative active-learning activities in physics. *Cognition and Instruction, 27*(2), 147–174. doi: 10/ckc7qw.

Shankar, R. (2015). *Physics 200 – Professor Shankar.* https://campuspress.yale.edu/rshankar/files/2015/09/shankarquotes-1ufgofr.doc. Accessed 4 Jan 2017.

Söderlund, H. (2016). *"Jättekul att det är så många tjejer här ikväll": En interaktionell studie om humor och kön i tv-programmet Parlamentet.* PhD thesis. Umeå University. http://urn.kb.se/resolve?urn=urn:nbn:se:umu:diva-120169.

Traweek, S. (1988). *Beamtimes and lifetimes: The world of high energy physicists.* Cambridge, MA: Harvard University Press.

Willey, A., & Subramaniam, B. (2017). Inside the social world of asocials: White nerd masculinity, science, and the politics of reverent disdain. *Feminist Studies, 43*(1), 13–41. doi: 10/gctj9r.

Chapter 7
"Significant Matter" in Sociomaterial Analysis of Educational Choices

Marianne Løken and Margareta Serder

7.1 Gaining New Insights into Educational Choices

"Something happens when the human and the material aspects meet that is important to understand", says physicist Karen Barad (1999, p. 2). Barad advocates including the physical, material reality when analysing phenomena that we normally tend to interpret as social and psychological. We term this a sociomaterial perspective (Løken and Serder 2018). In this chapter, we take up this perspective in order to better understand the reasons why three particular women chose to study STEM subjects, and consider the significance of *materiality* in their descriptions about their choices.

This study (which is taken from a more extensive qualitative study of atypical educational choices) explores how a sociomaterial perspective can provide a broader understanding of women's educational choices in general – and atypical educational choices in particular. The work demonstrates how gender and materiality are woven together in ways that are not generally revealed in sociocultural research about gender and educational choices. The chapter provides a necessary perspective to the debate on the recruitment of women to physics/STEM through its assertion that sociomaterial gender perspectives provide new angles for understanding educational choices.[1]

[1] The chapter is based on an article which forms part of a doctoral thesis (Løken 2017). The article is written in Norwegian and was published in NorDiNa in 2018.

M. Løken (✉)
University of Oslo, Oslo, Norway

M. Serder
Malmö University, Malmö, Sweden
e-mail: margareta.serder@mau.se

A. J. Gonsalves, A. T. Danielsson (eds.), *Physics Education and Gender*, Cultural Studies of Science Education 19,
https://doi.org/10.1007/978-3-030-41933-2_7

Despite decades of strategies to even out gender differences men still dominate in many of the sciences, including physics (NOU 2012, p. 15). Many attempts to understand the stubbornness of the gender divide in science look either at exclusion mechanisms or at individual choices. Researchers frequently explain the gender imbalance by reinforcing gender-stereotyped differences in interests, or by blaming gendered attitudes in the sciences (Adolfsson et al. 2011; Björkholm 2010). Gendered attitudes and differences in interests are also often discussed as being inherent to society and culture (Regan and De Witt 2015). In this chapter, we argue that seeing the gendering of educational choices as something that is constructed by sociomaterial practices and experiences provides greater opportunities for insight into how to recruit girls to STEM subjects, and to physics in particular. This view posits materiality as constitutive of everyday life, inherently entagled with its social dimensions. As Barad (2003) relates, "matter matters" and this obliges us as education researchers to consider the effects of materiality alongside considerations of language, discourse and culture.

When we examine educational choices as being situated, sociomaterial practices, we include people's relations to materiality in order to better understand *how* their interests and choices are formed. Sociomaterial approaches emphasise relationships that are established between human and *non*-human actors (Serder 2015). In physics, (one of the subjects of focus in this chapter) matter is inextricably linked to the subject both in practice and on a symbolic level. Not only is matter the research object of physics, physicists are symbolically bound to matter by its materials and artefacts: Atoms and sound are examples of materiality, but also physicists' laboratories, clothes and bodies. It is therefore natural to point to an interest in materiality as a potential reason for choosing a physics education. By stressing the importance of materiality, and of the interaction between human and non-human actors, these sociomaterial perspectives must be examined as a contributing factor when someone chooses their educational path.

In this chapter, we extend the argument to include materiality in the examination of psychosocial and sociocultural processes to interpretations of gender as developed in the field of Science and Technology Studies (Asberg and Lykke 2010; Fox et al. 2006; Lie 2003; Wajcman 2007). We highlight this connection to argue that analyses that emphasise the influence of sociomateriality on educational choices can generate new perspectives on the recruitment and inclusion of women in physics. In this chapter, we develop this argument by discussing empirical examples from three narratives taken from a larger qualitative study into women's atypical educational choices in Norway (Løken 2017). From close readings of these women's stories we developed the following questions: Which experiences and socio-material practices do these women deem to be relevant when deciding on an education? How does an atypical educational choice begin to take shape, and of what significance is gender in this process? For context on the importantance of adopting a sociomaterial gender perspective on educational choices, we begin by examining recent research into educational choices in Norway and similar Nordic countries.

7.2 Interests, Identity and Gender

Studies about educational choices carried out in Norway and in comparable Nordic nations often identify one's personal interest in a subject as a significant factor (Jensen and Henriksen 2015) in one's educational choices. Personal interest as a driving force in educational choice is reflected visually in recruitment campaigns, in the rhetoric used in information materials, and is evident in education policies. Indeed, one's choice of study in education can be said to represent one of the most dominant narratives in the discourse concerning educational choices (e.g. Bøe 2012; Løken 2015; Sinnes and Løken 2014). Additional research surveying recruitment to the sciences has found that personal interest, choice, and participation in the sciences are also closely linked to identity construction and socioeconomic background (Hazari et al. 2010; Holmegaard et al. 2014; Schreiner and Sjøberg 2007).

In Norway and other Nordic countries research demonstrates that girls more often choose career tracks related to biology or medicine, rather than to technology or physics – because (or partly because) these choices are more in line with their desired identities (Schreiner and Sjøberg 2007). Recruitment campaigns and products are designed with explicit connotations to femininity with a view to "selling" the message or product to women (Lagesen 2005; Lie 2003). However, a systematic review of the effects of management tools on influencing educational choices has concluded that it is difficult to isolate measures that boost the proportion of women in the sciences (Damvad 2015, 2016). The literature on educational choices includes research results that suggest girls feel alienated by the sciences, but also that such notions can be challenged (Løken 2015). Young women admit to many of the stereotypes that exist in the field, but also contest these in the narratives they provided for this study on their educational choices (Henriksen, Dillon and Ryder 2015; Løken 2015). As gendered recruitment campaigns appear to have a limited effect in terms of changing practices and opinions, we read our study participants' responses as a persuasive argument for exploring new perspectives on educational choice. Could there not be other ways of thinking about how a choice of education is shaped and gendered that could researchers with a deeper understanding of subject interest and identity construction? In this chapter, rather than try to find explanations for educational choices, we seek to deepen the understanding of how personal interests are constituted in sociomaterial practices.

The dominant perspective in the field of didactic research is the sociocultural perspective which is primarily about the interaction *between* human beings, and artefacts are primarily viewed as mediating instruments and not as active agents (Lenz Taguchi 2012). As mentioned earlier, the sociomaterial stance stresses that there is more to these interactions than mediation. In the next section we will provide a more detailed account of how a sociomaterial perspective can be used to identify experiences and practices that are relevant to the educational choices we make.

7.3 Educational Choices as Sociomaterial, Situated Practices

In post-humanist theory (e.g. Alaimo and Hekman 2008; Barad 2007; Solbrække 2011) materiality is used as an umbrella term for non-human actors. In short, it includes anything from technology, machines, artefacts, animals, texts and objects, or bodies and natural phenomena. Human actions are seen as mutually dependent on the material and social contexts in which they take place (Lenz Taguchi 2012), or as Tobias Roehl (2012) states: "Human actors and material objects are closely interwoven and transform each other in socio-material practice" (p. 110). To explain why educational choice can be identified as a sociomaterial practice it is relevant to study which material and human elements appear in individual stories about educational choices.

The significance of the material aspect to social life should not be taken to mean that "things" have agency in the sense of intention, but that the interweaving of the human being and the material object "does something" which in combination becomes something more than just the sum of the two. In sociocultural theory, objects, often referred to as artefacts, are seen as tools for human action. The sociomaterial approach instead emphasises how the presence of things and their availability determine what we can do – and how (Pickering 1995). According to Pickering (1995) human actions are the effects of the human and non-human actors' constant resistance and accommodation to each other, and to the sociomaterial conditions surrounding them. "Things" and "practices" can therefore be considered two sides of the same coin. Barad describes the output from the encounter with materiality as *intra-action*, and in this intra-action agency appears. Building on the works of Judith Butler, Michel Foucault, Bruno Latour and Donna Haraway (Hekman 2010), Barad writes that *"agential intra-actions are specific causal material enactments that may or may not involve humans"*, and that *"the world is intra-activity in its differential mattering*" (Barad 2003, p. 817). A sociomaterial approach directs the attention to what emerges from the encounter, the relationship entangling the human and non-human. We could therefore envisage a choice of education as an effect of such interweaving.

According to Ninni Sandvik (2015) the post-humanist question about which social and material relations link up to spark an action is an empirical one. A post-humanist, sociomaterial approach to empirical material challenges the way we understand actions, such as choosing an education. This does not necessarily mean that (all) relations are equal but that "it is not entirely clear who and what initiates, controls, prevents and is of significance" (2015, p. 52). Sandvik also suggests that neither is it clear "how the events occur or with which force the actors' agents for action negotiate for influence" (p. 52). Having acknowledged this, we will use a sociomaterial approach to examine *how* material experiences and practices have an impact on educational choices.

When one shifts from a sociocultural perspective to a sociomaterial perspective our understanding of the role of gender in educational choice changes. In both approaches gender is viewed as a practice and not as a distribution pattern of characteristics, a background variable or a biological or identity category. As Donna Haraway (2004) relates: "Gender is a verb. Not a noun. Gender is always about the production of subjects in relation to other subjects, and in relation to artifacts" (p. 329). In other words, gender is seen as something which involves *doing* and which is produced through relationships with the social and material world. In the following section, we use examples from our participants' narratives to demonstrate how gender gains agency in the sociomaterial experiences and practices referred to in their stories. We suggest that happens as participants constructed their identities and develop subject interests. This strategy offers a more nuanced view of the relationship between educational choice and gender, which in turn affects how we think about the recruitment of women and their inclusion in the sciences.

This research suggests that a sociomaterial perspective on educational choices recognizes the ways material experiences and practices construct or form *part* of Norwegian women's educational choices. The analysis of the empirical material does not seek to interpret descriptions of material experiences and practices as representations or expressions of subjective interests. Rather the point is to use intra-action as an analytical tool to investigate what is being produced by these experiences and practices. Thus, we are able to elaborate on the open empirical questions we formulated at the beginning of this chapter about: the ways an atypical educational choice takes shape; which experiences and practices are seen as relevant; and the significance of gender as a starting point for the analysis.

7.4 Narrative Analysis of Stories about Educational Choices as Material Practices and Experiences

Narrative analysis is an interpretive, hermeneutic methodology based on a review of qualitative data (Dauite and Lightfoot 2004). The empirical material that forms the basis for the analysis in this chapter is comprised of texts that describe Norwegian women's personal stories about their educational choices. The stories were collected over a three-year period (2009–2012) by the first author of this chapter (Løken 2017). The accounts were written in 2009 by 17 girls aged between 18 and 22 who entered higher education science programs with few female students, in the autumn of 2008 or 2009. The participants were all in the same age group and shared certain social patterns typical of their generation (Almås 1997). The informants were recruited through an open invitation. The invitation to participate in the study was published on a website, and the link was sent to Norwegian universities and university colleges where women were under-represented in science programmes. The young women were asked about: what or who inspired them to choose STEM; whether particular experiences, persons or other factors influenced their choice; how they felt about being one of just a few girls in their chosen study program; what they expected from their time as a student; their thoughts on future job and career prospects, etc. The website also contained information about the study's selection and participation criteria, information about how their stories would be used, and details concerning consent and anonymity. The researchers consciously opted to take what the informants had written at face value and did not conduct further investigations into the girls' backgrounds and social lives. The point was to study the factors that the informants themselves chose to present as being relevant to their educational choices.

A follow-up interview with the informants was carried out via email in 2011 and 2012. In the two follow-up interviews the participants were asked to reflect on stories about girls in the sciences, about whether recruitment drives such as bonus admission credits for girls have had any impact, about whether their expectations and study plans had changed, and about any advice they would give to young women wanting to study the same subject. The women were also asked why they think so few girls choose to study STEM subjects, about their take on gendered subject interests and educational choices, whether they would have chosen differently had they been born a boy, and more generally, their experiences as young women in a male-dominated academic environment.

The 17 submitted texts (130 pages) and transcriptions of the follow-up interviews were uploaded to the data analysis programme NVivo to structure and categorise the content. Examples of categories included: interests, identity, experiences and expectations. These categories provided a starting point for our theme-centred analyses of the empirical material as a whole (Thagaard 1998). All mentions of" material experiences," such as references to objects, sensations/emotions and relations to the non-human world, were thematised in the narratives. That very category formed the basis for the selection of the three stories used in the empirical analysis in this chapter.

Based on our interpretations of the material experiences in the preliminary analysis, we created condensed portraits of three stories. This reduction was an analytical strategy that helped elucidate and illustrate key aspects relevant to the problem posed in the chapter. Names and places cited in the excerpts were changed to ensure anonymity. We used excerpts from the accounts of Violet and Mia, who both studied physics at Norwegian higher education institutions, and also from Mona, who studied technical cybernetics at the university. Each account, in its own way, describes their educational choices as sociomaterial practices.

The analysis presented in this chapter examines the tension between a deconstructionist framework (in which the human being is seen to be positioned in and through competing discourses on the one hand) and a humanistic framework (in which the being's integrity is considered both the start and end point for the analysis). The deconstructionist framework allowed us to critique the idea of meaning as rational and straightforward. At the same time we strove to maintain a close eye on the significance of the material aspect in the analyses without giving preference to either the human subject or the material objects. The analysis of the three narratives did not allow for generalisation on statistical grounds. The point was, rather, to ascertain whether the interpretations we made could add more nuanced meaning to previous knowledge of young women's atypical educational choices, thus giving the reader new knowledge and insight.

7.5 Objectification and Embodiment in Narratives about Educational Choices

In the following analysis we look at the informants' observations about issues that often play a part in the recruitment of girls to the sciences: past experiences of science and of affectivity, technology in childhood, special treatment and school experiences – and we demonstrate how we interpret these themes using a sociomaterial approach.

The Sciences and Affectivity The sociomaterial approach to our data suggested that rather than look at past experiences of science or the conditioning of the informants' interest in science, we could interpret such experiences as sociomaterial *intra-actions*: experiences that could be linked to material and bodily aspects. While "interest" could be observed analytically through self-reports in questionnaires or interviews with informants and be reported as a thing in itself ("an interest"), the challenge for us became how to grasp signs of intra-actions, and how to understand what they had changed. Interest, in this analysis, was understood to be a transformation due to non-human (or sometimes human) agency.

In their responses, the informants suggested that playing with objects, building things, and solving tasks were an important factor that drove their interest in science. For example, Mia "couldn't get enough maths exercises in primary school," and described herself as someone who "loved playing with Lego, building things by following the manual, building playhouses and playing with toy cars." When we interpreted Mia's story in a sociomaterial perspective, the significance of materiality became conspicuous. Mia described how the challenges she confronted as a child constituted a significant part of the objectified experiences that she linked to her later choice of education. These challenges and their achievements, in turn, resulted in a sensation of excitement that she still remembered: "Especially as a child, it is very exciting to be set new challenges as a result of having achieved something," she wrote, also referring to how important it was to be set "additional exercises and extra material to work with." If, through a sociomaterial lens, we see objects as having agency, we can understand these experiences as moments where these materials and tasks moved something *within* Mia. Not only did she enjoy interacting with them, but they acted upon, and in some sense, changed her. This is why we interpreted this experience as an intra-action. In another example, Violet, responding to a question about an experience that motivated her educational choice, recalled a school trip in which the students were to build and launch a rocket.

Violet suggested that her experience of building a rocket launch on the school trip was a key factor in her motivation to study physics. It is not hard to imagine the thrill she experienced in this event—in being invited to connect with space itself. Taking this perspective in the analysis opened up the notion that the rocket, the space – or the place where this launch was taking place, changed Violet. Not only did she interact with the material while building the rocket, but the material world acted upon her. Do we know this for sure? No. Just as we cannot be certain about how the social or the cultural shape actions, we cannot know how the material informs them. However, to draw on Barad (2003), we cannot deny that "matter matters." In this analysis we suggest, therefore, that Violet's building of the rocket to launch was an experience wherein non-humans shaped humans, as an intra-action. The scientific artefacts' encounter with Violet sowed a seed that later informed her choice to study science.

When Mia and Violent described the factors that made them take an early interest in science, they spoke of their emotional experiences with material content. While an analysis from a sociocultural perspective would have stressed signs of (human) socialisation and enculturation in Mia's story, our sociomaterial approach instead focused on the agency of non-humans (e.g. the Lego bricks). A sociocultural analysis of our data would have put human interaction with the objects in the fore and focus, for instance, on the ways a child mimicks adults' actions in their interaction with these objects. However, if we interpret the Lego bricks as something that the informant forged important sociomaterial connections with (instead of, for example, attributing her interest in them to her parents' intentions and/or socioeconomic background) we can understand that the exposure to the material itself was an important experience for her.

Using the two examples from Mia's and Violet's stories about their educational choices, we have argued that the Lego bricks and rocket equipment had agency in the sense that they helped forged connections between the material and the human subject, and that these interactions were important enough to be mentioned in their women's narratives. A didactic consequence of these events suggests that such connections could be encouraged by exposing girls to material objects in different educational settings. While girls' educational choices are traditionally understood to be a consequence of their socioeconomic status, acknowledging the significance of affording them with material experiences can help researchers and policy makers move past an impossible problem: that schools cannot change their students' socioeconomic status.

7.6 Technology in Childhood

Technology is often defined as a masculine subject in literature, where the feminine and the masculine are pitted against each other (Lie 2003). To resolve this problem, technology is often reconstructed in feminine ways in order to encourage women to feel at home with the subject (Lie 2003.). These perspectives can also be found in the literature addressing so-called "girl-friendly" initiatives (Sinnes and Løken 2014). Donna Haraway (1991) challenges this viewpoint, stressing how technology has become an important part of our lives. By examining how the human subject intra-acts with technologies, we therefore also need to deconstruct the link between technology and masculinity. By looking at how materiality is given agency in our informants narratives, we can understand the subject matter as part of the subject's lived experience without cultural dichotomies dictating how gender plays a part in the educational choice.

The next example details Mona's material experiences with technology. Mona grew up in an industrial town and attended the company crèche of the biggest local employer. Here, she often heard about the rocket-making factory: "Big influences have been the place I grew up and my family. Where I come from there are a lot of civil engineers working for the big tech companies [...] These businesses specialise in weapons systems, marine technology, defence technology, offshore technology, gas turbines, car parts, oil and gas and aviation [...] And these companies are highly noticeable in the local community since they sponsor sports clubs, arts projects and events, and they donate technology aids to schools and hold various family days, open days and school trips where you get to see what they do. Through the years I've been on many such days, and since several members of my family work for some of these companies, I've got an insight into the technologies they develop. That has always interested me [...] I have also had the opportunity to work in several of the companies. This has only been part-time jobs." These experiences gave Mona a "...positive impression of those who work there, the social environment and what they work with." From a sociomaterial approach, Mona's experiences can be viewed as an example of intra-action between the human and material – her play with technology and the practical experiences that the local employer represented. Her story excerpts reveal material experiences and practices that have been present since her childhood. Seeing, hearing about and experiencing a highly technological world filled with material artifacts captured Mona's imagination. Remembering these objects and the excitement they offered remain with her.

What can we glean from this memory? We suggest that Mona's material experiences tell us something about the importance of being a participating actor in an environment that has an expansive technological repertoire. However, the different technologies she described require knowledge of how to use them. Technology is, therefore, a combination of objects, practice and knowledge that Mona encountered in various settings throughout her childhood. The narrative demonstrated how the technology she grew up with contributed to her identity construction and influenced her decision to study technology.

7.7 From the Significant Other to Significant Matter

In the study, the informants described themselves as being free, independent decision-makers who ultimately act in accordance with their subject interests and aspirations. By "degendering" their bodies, presenting them as gender-neutral in their descriptions, distancing themselves from symbols and identity markers associated with the female and feminine, describing themselves as "tomboys" who have "always" been interested in science they took control of their bodies so that being/having a female body did not come into conflict with their educational choices. In these responses we see these young women distancing themselves from cultural conceptions of womanhood by rejecting feminine norms and values and by choosing an educational path that demonstrates more traditionally masculine ideals. In this way, we suggest, they create opportunities for practising gender outside stereotypical gender discourses. We assert that, even though the informants identified opportunities and limitations that directed their choices, they were deeply rooted in the material world. In the final section we highlight some implications of using a sociomaterial, theoretical approach when examining educational choice based on the insights we have gained through our analyses. But first, a few more reflections on how material experiences and practices have an impact on educational choices.

To aid in our empirical analysis, we asked the informants questions about how material experiences and practices intra-act, and about how body and gender represent materiality in their educational choices. To arrive at our study's conclusion, we have studied narratives about educational choices to identify what is given agency in the narratives. By interpreting stories of Norwegian women who have opted to study science in a male-dominated field, we have argued that a sociomaterial perspective (following Barad's (2007) concept of intra-action) can be used to understand how material experiences and practices may influence educational choices. We suggest that it may be useful to pursue this approach in order to gain a broader understanding of gendered educational choices. It might also be useful to analyse other kinds of data about women's educational choices from this analytical point of view. In a sociomaterial perspective gender does not serve as a cause but is negotiated as part of one's educational choice.

Rather than dismiss educational choices as the result of an intentional choice according to interests or cognitive, rational processes and a consequence of socioeconomic background, we have also shown that material forms of agency come into play when women who have chosen a science in which they are a minority tell their stories. Our interpretations of the stories about their choices reveal how embodied, sensory experiences and intra-actions with concrete things are understood as key components in the formation of the decision to study science: in Mia's case her experiences with building blocks and construction games in childhood were key, while Violet pointed to factbooks and films as important artefacts in the story about her choice, in addition to recounting her material experiences of rocket-building and launching. In Mona's narrative her encounter with technology in the local community came to have a major impact on her interests and choices. From these stories we see that the things that surround us, the experiences we have with them, and our bodily situation in the world critically inform our interests, aspirations, and educational choices.

In a sociomaterial approach, the human species is not considered to be superior to other subjects, nor is it believed to be the only subject endowed with agency. This means that the material object has a natural and equal place on par with the human subject in analyses of complex, social processes and practices – such as educational choices. According to Lenz Taguchi (2012), a sociomaterial approach to learning enables us to create a language that better describes the significance of the material aspect, because this approach involves utilising differences, diversity and complexity in relation to learning. We believe that this is valuable knowledge when it comes to studying educational choices. It means that learning – such as choosing an education – becomes a bodily and material question along the same lines as questions about thinking and cognition based on language and discourse (Lenz Taguchi 2012).

Estrid Sørensen (2009) argues that in education research there "is a blindness toward the question of how educational practice is affected by materials" (p. 2). We believe the same can be said about our understanding of educational choices as a sociomaterial practice. These are perspectives that challenge the ideas about the consequences of material forms of agency (Barad 1998). In a discourse about educational choices this can be taken to mean that the material objects we surround ourselves with form part of those choices – and that the subject is undergoing a continual, constitutive process (Braidotti 1994). Thus the informants' continual identity construction is closely interwoven with the materials in their surroundings—what we identify as "significant matter" in the stories about educational choice. Existing research has long identified "the significant other," such as a teacher, family member or peer, and shown how they can all exert influence over educational choices (Danielsson 2013; Sjaastad 2011). In this chapter we have demonstrated how material experiences and practices also impact educational choice, and suggest that this non-human and material aspect, the "significant matter" must also be accounted for in further studies on educational choices.

7.8 Educational Choices as Sociomaterial Intra-Actions

This chapter can be read as an argument for the interpretation of educational choice as a sociomaterial intra-action (Barad 2007). This analysis both challenges and supplements the more traditional sociocultural approaches to choosing an education in which the decision is seen primarily as an individual choice based on interests formed by the social environment. The key difference between the sociocultural and sociomaterial approaches is how the human subject is understood and positioned. As we have shown, a sociomaterial analysis implies that the human subject is no longer at the centre of the analysis (Løken and Serder 2018). We conclude with the assertion that "significant matter" must be included when analysing educational choices on a level footing with "the significant other." And if we want to influence women's educational choices through different forms of practice such as recruitment drives, teaching practices and careers advice/guidance, we would do well to explore how material practices and experiences impact individual choices. Or, using post-humanist terminology, we must acknowledge that the material object "kicks back"(Løken and Oyselbø Sørensen 2018) in studies on educational choices, in teaching, and in initiatives aimed at recruiting more girls to the sciences.

References

Adolfsson, L., Benckert, S., & Wiberg, M. (2011). Gapet har minskat: skillnader mellan högoch lågpresterande flickors och pojkars attityder till biologi, fysik och kemi 1995 och 2007. *NorDiNa, 7*(1), 3–16.

Alaimo, S., & Hekman, S. (2008). *Material feminism*. Bloomington/Indianapolis: Indiana University Press.

Almås, R. (1997). Tre generasjoner rekonstruerer sin ungdom. In Frønes (red.) *Livsløp – oppvekst, generasjon og sosial endring*. Oslo: Universitetsforlaget.

Asberg, C., & Lykke, N. (2010). Feminist technoscience studies. *European Journal of Women's Studies, 17*(4), 299–305.

Barad, K. (1998). Getting real: Technoscientific practices and the materialization of reality. *Differences: A Journal of Feminist Cultural Studies, 10*(2), 87–128.

Barad, K. (1999). Agential realism: Feminist interventions in understanding scientific practices. In M. Biagioli (Ed.), *The science studies reader*. London: Routledge.

Barad, K. (2003). Posthumanist performativity: Toward an understanding of how matter comes to matter. *Signs: Journal of Women in Culture and Society, 28*(3), 801–831.

Barad, K. (2007). *Meeting the universe halfway. Quantum physics and the entanglement of matter and meaning*. Durham og/London: Duke University Press.

Björkholm, E. (2010). Technology education in elementary school: Boys'and girls´ interests and attitudes. *NorDiNa, 6*(1), 33–42.

Bøe, M. V. (2012). *What's in it for me? Norwegian students' choices of post-compulsory science in an expectancy-value perspective*. Doktorgradsavhandling, Universitetet i Oslo.

Braidotti, R. (1994). *Nomadic subjects. Embodiment and sexual difference in contemporary feminist theory*. New York: Columbia University Press.

Damvad Analytics. (2015). Styringsvirkemidler som påvirker utdanningsvalg. Kunnskapsoppsumering og analyse. *Rapport* 02/06/15.

Damvad Analytics. (2016). Piger i science, technology, engineering and mathematics (STEM). Kortlægning af utfordringer inden for køn, ligestilling og udannelse i Norden. *Rapport* 18/01/16.

Danielsson, A. T. (2013). Science for whom? Case studies of two male primary school student teachers'construction of themselves as teachers of science. *NorDiNa, 9*(2), 145–155.

Dauite, C., & Lightfoot, C. (2004). *Narrative analysis. Studying the development of individuals in society*. Thousand Oaks: Sage.

Fox, M. F., Johnson, D. G., & Rosser, S. V. (2006). *Women, gender and technology*. Urbana/Chicago: Illinois University Press.

Haraway, D. (1991). *Simians, cyborgs, and women: The reinvention of nature*. London: Free Associations Books.

Haraway, D. (2004). *Modest_witness@second_millenium. The Haraway reader*. New York/London: Routledge.

Hazari, Z., Sonnert, G., Sadler, P. M., & Shanahan, M. C. (2010). Connecting high school physics experiences, outcome expectations, physics identity, and physics career choice: A gender study. *Journal of Research in Science Teaching, 47*(8), 978–1003.

Hekman, S. (2010). *The material of knowledge. Feminist disclosures*. Bloomington/Indiana: Indiana University Press.

Henriksen, E., Dillon, J., & Ryder, J. (2015). *Understanding student participation and choice in science and technology education*. New York/London: Springer Dordrecht Heidelberg.

Holmegaard, H. T., Ulriksen, L. M., & Madsen, L. M. (2014). The process of choosing what to study: A longitudinal study of upper secondary students' identity work when choosing higher education. *Scandinavian Journal of Educational Research, 58*(1), 21–40.

Jensen, F., & Henriksen, E. K. (2015). Short stories of educational choice – In the words of science and technology students. In E. K. Henriksen, J. Dillon, & J. Ryder (Eds.), *Understanding*

student participation and choice in science and technology education (pp. 135–151). Dordrecht: Springer.
Lagesen, V. (2005). Fra firkanter til rundinger? produksjon av femiistisk teknologipolitikki en kampanje for å rekruttere jenter til datastudier. *Kvinneforskning* 1.
Lenz Taguchi, H. (2012). *Pedagogisk documentation som aktiv agent: Introduktion til intra-aktiv pedagogic*. Malmö: Gleerups.
Lie, M. (2003). He, she and IT revisited. *New perspectives on gender in the information Society.* Oslo: Gyldendal Akademisk.
Løken, M. (2015). When research challenges gender stereotypes: Exploring narratives of girls' educational choices. In E. K. Henriksen, J. Dillon, & J. Ryder (Eds.), *Understanding student participation and choice in science and technology education* (pp. 277–295). Dordrecht: Springer.
Løken, M. (2017). *Skriv ditt valg! Nyskriving av historier om @typiske utdannings-valg*. Doktorgradsavhandling, ph.d., Det matematisk-naturvitenskapelige fakultet, Universitetet i Oslo.
Løken, M., & Oyselbø Sørensen, S. (2018). Materielle praksiser ogerfaringer "kick back". En sosiomateriell analyse av beretninger om utdanningsvalg. *NorDiNa, 4*(4), 366–378.
Løken M. & Serder M. (2018) In-between chapter: Troubling the social – Entanglement, agency, and the body in science education. In Otrel-Cass K., Sillasen M., Orlander A. (red.) *Cultural, social, and political perspectives in science education* (pp. 133–137). Cultural Studies of Science Education, 15. Springer, Cham. doi:https://doi.org/10.1007/978-3-319-61191-4_11.
NOU 2012:15. (2015). *Politikk for likestilling*. Oslo: Departementenes servicesenter.
Pickering, A. (1995). *The mangle of practice: Time, agency, and science*. Chicago: University of Chicago Press.
Regan, E., & De Witt, J. (2015). Attitudes, interest and factors influencing STEM enrolment behaviour: An overview of relevant literature. In E. K. Henriksen, J. Dillon, & J. Ryder (Eds.), *Understanding student participation and choice in science and technology education* (pp. 63–88). Dordrecht: Springer.
Roehl, T. (2012). Disassembling the classroom – And ethnographic approach to the materiality of education. *Ethnography and Education, 7*(1), 109–126.
Sandvik, N. (2015). Posthumanistiske perspektiver. In A. M. Otterstad og A. B. Reinertsen (Red.), *Metodefestival og øyeblikksrealisme – ekseperimenterende kvalitative forskningspassasjer* (pp. 45–62). Bergen: Fagbokforlaget.
Schreiner, C., & Sjøberg, S. (2007). Science education and youth's identity construction – Two incompatible projects? In D. Corrigan, J. Dillon, & R. Gunstone (Eds.), *The re-emergence of values in science education* (pp. 231–247). Rotterdam: Sense Publishers.
Serder, M. (2015). *Möten med PISA. Kunskapsmätning som samspel mellan elever och provuppgifter i och om naturvetenskap*. Malmö: Malmö Høgskola.
Sinnes, A., & Løken, M. (2014). Gendered education in a gendered world: Looking beyond cosmetic solutions to the gender gap in science. *Cultural Studies of Science Education, 1*(1), 343–364. Springer.
Sjaastad, J. (2011). Sources of inspiration: The role of significant persons in young people's choice of science in higher education. *International Journal of Science Education, 33*, 1–22.
Solbrække, K. N. (2011). Maskulin (u)orden i norsk sykepleieutdanning. I Leseth, A. og Solbrække, K. N. (red.), *Profesjon, Kjønn og Etnisitet*. (pp. 35–55). Latvia: Cappelen Damm AS.
Sørensen, E. (2009). *The materiality of learning: Technology and knowledge in rducational practice*. Cambridge and New York: Cambridge University Press.
Thagaard, T. (1998). Systematikk og innlevelse. En innføring i kvalitativ metode. Fagbokforlaget.
Wajcman, J. (2007). From women and technology to gendered technoscience. *Information, Communication & Society, 10*(3), 287–298.

Chapter 8
Disability in Physics: Learning from Binary Mistakes

Adrienne Traxler and Jennifer Blue

8.1 Introduction

Physics has its own particular stories that allow for exceptional models of diversity. Many young scientists have been inspired by Marie Curie, or recognized a role model with visible mobility impairments in Stephen Hawking. But these stories tend to be fused with the ideal of brilliance: if your mind is strong and pure and unique enough, you can transcend your limitations (of being a woman, of being in a wheelchair). In truth, all kinds of people deserve the opportunity to become physicists, and physics deserves the contributions of all kinds of people. To build a physics that truly welcomes the talents of diverse individuals, we must learn to tell better stories.

8.2 Disabilities in STEM Higher Education

Work elsewhere has reviewed some of the ways that gender can signal "not-belonging" in physics and STEM (Faulkner 2009; Gonsalves 2014; Hill et al. 2010; Ong et al. 2011), and intersectional studies in this work remind us that gender is never the whole story. In this chapter, we first turn to the disability studies literature for foundational themes and a few STEM-specific points.

A. Traxler (✉)
Wright State University, Dayton, OH, USA
e-mail: adrienne.traxler@wright.edu

J. Blue
Miami University, Oxford, OH, USA
e-mail: bluejm@MiamiOH.edu

A. J. Gonsalves, A. T. Danielsson (eds.), *Physics Education and Gender*, Cultural Studies of Science Education 19,
https://doi.org/10.1007/978-3-030-41933-2_8

8.2.1 For Students

College and university students in the United States and the United Kingdom are invited to tell their institutions if they have any disabilities for which they need accommodations. Some students avoid this identifier, choosing not to "'accept, disclose, or document' a status of 'disabled student'" (Jacklin 2011, p. 99). Although colleges and universities are only required to provide accommodations for students who disclose disabilities, past experiences with discrimination or fears of future discrimination cause some students to stay quiet. Some also feel that accommodations such as extra testing time—even when essential to their success—give them an unfair advantage (James et al. 2018).

Students who study education may learn little about disabilities. Both academic journals and some academic departments reinforce the segregation of disability by keeping research about special education out of general education journals and spaces, as well as only having one or two courses on special education and disability for general education teacher preparation programs (Connor et al. 2016). This reinforces the "othering" of those with a disability, positioning them as a rare exception rather than an integral part of classrooms.

Undergraduates with disabilities study STEM at the same rate as undergraduates without disabilities: 11% of students with disabilities (Sevo 2012; Sutton 2017) and 12% of those without declare STEM majors (Sutton 2017). There are still differences: students with disabilities are more likely to enroll at a 2-year school than a 4-year school,[1] and they are less likely to pursue graduate study, which is often needed for careers in STEM (Sutton 2017). Note that these figures come from self-reports of current students; institutional census data reports much lower numbers of 1–5% of undergraduates with disabilities (National Science Foundation and National Center for Science and Engineering Statistics 2017, p. Technical Notes), and completion rates are not part of this comparison.

In order to be more welcoming to people with all sorts of disabilities, labs and classrooms need to be more physically accessible. For some STEM disciplines, outdoor spaces and public areas should also be accessible (Carabajal et al. 2017). There could be ramps in hiking areas, automatic doors, and the like. This would be a welcome change from institutional policies and lack of accommodations that keep people out (Marks 2017).

Students have trouble finding mentors with disabilities. They must contend with widespread ignorance, not only about their specific disabilities (whether visible or invisible) but about the Americans with Disabilities Act (Wilkie 2014). Even if someone who is Deaf/hard of hearing does have a sign language interpreter, the interpreter is unlikely to be equipped with STEM jargon (Wilkie 2014). When they

[1] In fact, students with disabilities at 2-year schools are more likely than other students at 2-year schools to pursue STEM (Lee 2014). A recent demographic review by Kanim and Cid (2017) shows that two-year colleges are seriously under-represented in physics education research studies. This bias adds to the invisibility of students with disabilities in physics.

finish their studies, people with disabilities are more likely to pursue STEM careers in the government and less likely than others to go into education. This has implications for visibility, as students are unlikely to know any disabled STEM teachers (Sutton 2017).

Graduate students across fields are at risk of mental health disorders. A 2015 study of graduate students at Berkeley found that nearly half of them had significant indicators of depression (The Graduate Assembly 2015); this was a follow-up to a 2005 study that had found that 10% of graduate students had seriously considered suicide (Berkeley Graduate and Professional Schools Mental Health Task Force 2004). An even more recent study found that PhD students in Belgium were at higher risk of mental health disorders (particularly depression) than other highly educated people (Levecque et al. 2017). A prevailing culture that graduate school is supposed to be stressful can make it more difficult for graduate students with disabilities to ask for or receive accommodations. This situation is likely to be exacerbated in departments with a weed-out culture, which is common in many prestigious physics programs. If campus disability resources are only designed with undergraduates in mind, graduate student accommodations may be handled within the department or according to faculty advisor notions of need, which are idiosyncratic and may not be compliant with the law. Indeed, many college faculty only partially understand how they are required to accommodate disabilities in their teaching and are often unfamiliar with campus resources for these students (Paul 2000).

Perhaps for these reasons, students with disabilities are less likely than their peers to earn graduate degrees in STEM. In 2014, about 6% of STEM doctorates in the United States were earned by students with disabilities (National Science Foundation and National Center for Science and Engineering Statistics 2017). About 6% of those STEM doctorates were in physics or astronomy, comparable between students with and without disabilities. Ultimately, only about 2% (Slaton 2013) to 7% (Sevo 2012) of the STEM workforce is made up of people with disabilities[2].

8.2.2 *For Faculty*

For faculty living with chronic illnesses and disabilities, the procedures are less clear than they are for students. Usually the office which provides accommodations works only with students, so faculty must work with multiple offices at their schools (if they can determine which office or offices to work with). A survey of faculty

[2] Even at this level of summary statistics, numbers are elusive because how (or if) the data is collected varies widely. The figures for PhDs come from the Survey of Earned Doctorates, which in 2005 reported less than 2% of STEM doctorates going to students with disabilities (Committee on Equal Opportunities in Science and Engineering 2009). In 2012, the question from which these numbers derive was changed from "Are you a person with a disability?" to ask about functional limitations in several areas, so numbers before and after 2012 are not directly comparable.

reported that fewer than half were familiar with accommodations that they may have been entitled to, and fewer than 15% had asked for accommodations (Price et al. 2017). Many faculty choose not to disclose their disabilities, particularly mental disabilities. Instead, they handle their accommodation needs informally or privately, if at all. This is particularly disturbing given a recent UK study that found that the percentage of academics with mental illness is over 50% (Wilcox 2014). For faculty who did not seek accommodations, fear of stigma was a recurring theme (Price et al. 2017). Specific concerns included workplace gossip, loss of credibility, or harming chances of promotion, tenure, and contract renewal. For faculty in untenured or short-term positions, or still on the job market, requesting accommodations is inherently risky (Adjunct 2008).

In the survey by Price and collaborators, the most common reason for not disclosing was "Feeling that it's not other peoples' business" (62% of non-disclosing respondents). For faculty as well as students, "the disabled person who is 'invisible' is responsible for making himself visible, or discernible. When we make this assumption, responsibility for alleviating injustice is placed upon the person suffering the injustice in the first place. Oppressed persons should not bear the burden of educating and reforming their oppressors, and yet, that is what the visible/invisible metaphor asks of disabled people" (Price et al. 2017). What makes this even more disturbing is that faculty who choose to remain invisible are unlikely to appear as role models and mentors for their students, although the need for both non-disabled and disabled people to see disability in STEM fields is crucial, both to the field they are in and society as whole (Marks 2017).

Compared to gender, or to race and ethnicity, students with disabilities are more equally represented by number in seeking undergraduate degrees. Without intersectional data, it is unclear whether these numbers are even across demographic groups (and it is likely they are not; see Lee 2014). At the graduate level, the situation is somewhat worse. For many STEM careers, an undergraduate degree is necessary but not sufficient, and the references above document some of the ways that students with disabilities are filtered away from some scientific careers.

8.3 Critical Frameworks of Disability

The next theoretical task we set for ourselves was challenging: to combine elements of several frameworks that are mostly used outside of science education, discuss them in a physics context, and distill themes to guide study of disability in physics. Below we introduce key elements of disability identity, critical race theory, DisCrit, and crip theory. We include these perspectives because one major lesson from our past work was how deeply embedded and implicit are the frameworks that all of us, as researchers and as humans, have about gender (Traxler et al. 2016). We have called it the binary deficit model: the idea that there are only two kinds of people, that the most important (useful, scientific) kind of study is to compare them on some quantitative measure, and that from this ranking one can plan how to fix the more deficient group. This may sound like a harsh exaggeration, and we do not claim that it is the conscious paradigm of any particular researcher. But it emerges in aggregate, from the overwhelming prevalence of binary, sex-coded, quantitative comparisons that use men's scores or performance as the standard to aspire to and that suggest remediations for women[3].

Sexism is not the same as racism; ableism is not the same as transphobia. We do not want to simplify the nuances of different struggles against oppression. However, it is possible to learn things from the study of one area that help us anticipate and recognize bias and discrimination when they occur for similar reasons in another place. So in trying to fuse a more inclusive understanding of how disability plays out in physics, we begin by listening to what researchers have said about other facets of identity. We start with key themes from the disability studies literature, focusing on what seems most relevant to science in higher education. Next we introduce some elements of critical race theory, a framework made for the purposes of deconstructing the far-reaching, bitterly entrenched implications of racism. In a field as improbably White as physics, it would be irresponsible not to draw on this expertise. The next piece discusses DisCrit, or dis/ability critical race theory, a recent synthesis of those two frameworks. We will touch on this in parallel with crip theory, which studies the intersection of disability with LGBT identities. This work has primarily appeared in literature or film criticism, with some extension in the medical field (Egner 2016; McRuer 2006b). The climate for LGBT physicists is often unfriendly or unsafe (Atherton et al. 2016), and we hope to learn from this area of scholarship as well.

This work is not expected to synthesize all the answers. Rather, we hope that by drawing in different threads of identity study, we can be more aware of the places where multiple identities can complicate or intensify issues faced by physicists with disabilities. We also aim to outline concepts and collect useful references for physics or science education researchers hoping to learn more.

8.3.1 Disability Identity

The question of what is defined as a disability is ongoing. A crucial first distinction to make is between medical and social models of disability. In the medical model, disability is "an individual condition and a problem needing medical solutions" (Linton 1998). It arises from biology and is primarily treatable by doctors' interventions, with the goal of fixing deficiencies in the disabled person (Egner 2016; Shakespeare 1996). The social model argues that people have impairments, but that a person is disabled by the societal conditions that surround them. A person who uses a wheelchair may not be disabled in surroundings with curb cuts, ramps,

[3] Work that tests interventions can make a real difference for women or other marginalized students (Brahmia 2008; Hill et al. 2010). But there is an often-neglected flip side: as Brickhouse and Potter (2001) put it, "what needed transformation was the schools, not the girls."

and elevators, but is disabled by a building that requires the use of staircases (Connor et al. 2016; Linton 1998).

The medical model underpins many legal and official documents. The Americans with Disabilities Act (ADA) defines a disability as "(i) A physical or mental impairment that substantially limits one or more of the major life activities of such individual; (ii) A record of such an impairment; or (iii) Being regarded as having such an impairment" (Americans with Disabilities Act Title II Regulations n.d.). Significantly, the individual involved may not be the one who defines themself as disabled; if there is a record, or others regard them as having an impairment, then they are disabled.

The ADA does not define a set of qualifying impairments, and there is no universal list. Table 8.1 shows two sample lists of disabilities. The first is from the Wright State University Office of Disability Services, the list that students see when they register themselves for accommodations (Wright State University Office of Disability Services 2018). The second is from the United States Department of Labor's "Voluntary Self-Identification of Disability Form," which collects data when people apply for jobs at which they may require accommodations (United States Department of Labor n.d.). These lists overlap, but they are not identical.

A person with impairments included under broader criteria might find themselves declared non-disabled by a more restrictive list. In either case, the authority to decide rests with the institution, often in communication with a doctor. This brings us back to the heart of the medical/social model contrast: the underlying structure of power and responsibility. The medical model casts disability as the problem of the individual, who must ask for accommodations and rely on the kindness of authority

Table 8.1 Two different lists of disabilities

Wright State University Office of Disability Services (WSU ODS 2018)	US Department of Labor "Voluntary Self-Identification of Disability Form" (US DOL n.d.)
ADD/ADHD	Blindness
Aspergers/Autism Spectrum Disorder	Deafness
Hearing Impairments	Cancer
Learning Disability	Diabetes
Other Health Impaired	Epilepsy
Other (multiple disability, deaf/blind)	Autism
Physical Disability	Cerebral palsy
Psychological/Mental Health	HIV/AIDS
Speech/Language Impairment	Schizophrenia
Traumatic Brain Injury	Muscular dystrophy
Visual Impairments/Blindness	Bipolar disorder
	Major depression
	Multiple sclerosis (MS)
	Missing limbs or partially missing limbs
	Post-traumatic stress disorder (PTSD)
	Obsessive compulsive disorder
	Impairments requiring the use of a wheelchair
	Intellectual disability (previously called mental retardation)

figures to provide them. The social model contends that disability is a social problem, which must be solved at that level by a fundamental shift in priorities to make institutions accessible to all (Burgstahler 2015).

"What is a disability" shifts depending on who is doing the defining and what the stakes are. Gee (2000) distinguishes several senses of "identity" that help to unpack this idea. The four ways he proposes of viewing identity are: natural (something you "are"), institutional (an assigned position), discourse (a way you are recognized by others), or affinity (a shared set of focused practices). The medical model might be said to rest on natural and institutional views of identity—a person "is disabled," with the implication of a static individual state, or "receives accommodations" because of a medical diagnosis filtered through university policies. The social model brings in discourse and affinity identities. How (or if) to acknowledge a disability to peers or mentors, and how they may react, lives in the realm of discourse identities. Groups such as the International Association for Geoscience Diversity (https://theiagd.org/) harness the power of affinity identities for collective action. All of the above senses are folded into the term "disability identity." To distinguish between them in a given situation or research study, we must keep these questions in our pockets: Who has the authority here? Who gets to decide?

8.3.2 Notes from Critical Race Theory

Critical race theory (CRT), which came out of the study of law, takes the position that racism is normal in American society, so ingrained that it now seems natural to White people. The law might be able to help with the most blatant and extreme racism, like explicit policies of discrimination against non-Whites, but the law cannot help with everyday microaggressions and more "acceptable" forms of racism (Delgado 1995; Delgado and Stefancic 2012; Essed 1990).

Embracing CRT means realizing that even the mainstream civil rights discourse still believes in the idea of a meritocracy, assuming that it will be possible to level the playing field and that people of all races will then have an equal chance to compete for what they need and want. Fully embracing CRT means realizing that this is not so; we need first to question why "jobs, wealth, education, and power are distributed as they are" (Crenshaw et al. 1995, p. xv). It may no longer be possible to use the courts to correct the injustices of racism without examining the role that the courts and the law itself has played in sanctioning racism. Law schools themselves are embedded in our racist culture, so lawyers who have succeeded there are likely to have racist attitudes themselves, albeit unconscious ones (Crenshaw et al. 1995).

An important facet of CRT is the rejection of the idea that research and scholarship can be objective or neutral; since everything we do is embedded in society, everything is political. Our education system positions affluent White voices as the "standard" knowledge (Ladson-Billings 1998), and physics is steeped in this tradition as well. Racism is ingrained in society, so research about race is either entrenched in its own racism or actively fighting against it (Crenshaw et al. 1995).

CRT urges us to fight this ingrained racism with counter-storytelling, rewriting narratives about people of different races. It is hoped that the actual stories of individual people of color will both show how and why they push for change and, eventually, change the norms of society (Delgado 1995; Delgado and Stefancic 2012; Rosa and Mensah 2016).

The idea of rejecting objectivity and neutrality pulls at one of the central tensions in physics education research. Physics has thrived as a science of reductionism, with the objective search for truth as a core value (Whitten 1996, 2012). Appeals to objectivity have historically distinguished science from non-science (Gieryn 1983; Harding 1986), so stepping away from this ideal can threaten a researcher's place in a physics department.[4] This element of CRT aligns with critiques of objectivity in feminist science studies (Harding 1986) and with calls for emancipatory research to break down traditional divides between researchers and subjects (Liasidou 2014; M. Oliver 1992; Whitten 2012). It also calls for pushing against decades—arguably centuries—of disciplinary norms, so it is no surprise that PER has not seen a wild proliferation of critical (in the CRT sense) perspectives.

8.3.3 DisCrit and Crip Theory

"DisCrit" is a crossover field, grown from combining disability studies and critical race studies. Guiding tenets include (1) recognizing that while categories such as race and disability are socially constructed, they do have material and psychological consequences, (2) valuing multidimensional identities, and (3) amplifying voices and stories of people from marginalized groups (Annamma et al. 2016).

Both racism and ableism are deeply ingrained in our society. Both are considered normal to the dominant culture, and science (or pseudoscience) has been used as a tool to reinforce both of them (Dolmage 2017; Gould 1996). Racism and disability are both social constructs, and therefore both can change. People of color and disabled people suffer economic disadvantages as a result of discrimination, which prevents people in both groups from being able to fully participate in society (Connor et al. 2016).

Disabled people should not be treated like children; they can usually live independent lives without a guardian, they can advocate for themselves, and they can tell their own stories. These counternarratives are vital. Of course, intersectionality is important. Some traditional ways of activism (marches and other forms of physical civil disobediences) are often rooted in ableism, though this is not necessarily on purpose. People need to recognize the importance and need for diverse forms of resistance, which does not have to be physical in order to be worthy (Connor et al. 2016).

[4] Objectivity is a concept with its own complicated history (Daston 1992) and plethora of meanings (Barad 2007; Harding 1986), but we take as given here that it is held as a core value of modern physics.

DisCrit and crip theory both lay open how disability intersects with other facets of identity, in Crenshaw's sense of "intersection"—that people caught at these borders may be uniquely underserved by efforts made on behalf of a single identity "group." Efforts made to help women, for example, carry the organizers' ideas of what "women" are and what support they need. "Women in science" alliances do not typically drive efforts to establish safe and accessible bathroom use on campus. These efforts tend to be seen as an LGBT rights issue or possibly the domain of an office of disability services, even though they are a more pressing concern for many women than workshops on future salary negotiation.

No one advocacy group can do everything, of course—the point here is that as researchers, if we want to do something useful for our participants, we have to be aware of how people's identities are separated, sorted, and categorized by the social and institutional power structures that we are all embedded in. Earlier work has reviewed male/female binaries and how those have consciously or unconsciously shaped what research questions are asked in physics education, the methods used to probe them, and how results are framed (Danielsson 2010; Traxler et al. 2016). DisCrit and crip theory both bring out ways that disability intersects with other facets of identity, further complexifying this binary model.

Not all facets of identity have the same substance, which Gee (2000) explores in the context of education. Gee uses "being ADHD"[5] as an example to tease apart natural, institutional, discourse, and affinity views of identity—ways of being a certain kind of person. Many identity statements could start with "I am..." but "I am an identical twin" indicates a fairly straightforward natural identity, a fact of birth. "I am a physicist" includes elements of training, a job, recognition by peers—a blend of institutional, discourse, and affiliation identities. The more one differs from the expected presentation, the more elusive recognition can be.

DisCrit focuses much of its criticism on the first two of these categories, natural and institutional. Some DisCrit studies analyze how powerful racist stereotypes collide with officially diagnosed impairments—beliefs about natural and institutional identities—to produce overrepresentation by race in certain official categories of disability (Artiles et al. 2010; Connor et al. 2016). Crip theory has its origins in literary criticism, with a focus on discourse (Egner 2016; McRuer 2006b). In Gee's terminology, crip theory might be said to focus on institutional and discourse identities. A crip theory perspective might consider how a student is categorized by their university (as "needing accommodations," and what kind) and how this plays out in their daily life as they are recognized by peers and faculty. Though not explicitly crip theory, Slaton (2013) raises parallel issues of what bodies are regarded as "normal" in STEM laboratory courses. These assumptions about bodily normalcy[6] combine with other images of engineers' identity (White, straight, etc.) but also

[5] Attention-deficit/hyperactivity disorder, also previously known as attention deficit disorder.

[6] Defining and enforcing "normal" has become a society-wide endeavor linking prisons, factories, and schools (Foucault 1979). At universities, power is handed down by credentialed authority figures in elaborate ceremonies. This credentialization process becomes its own justification for maintaining authority (P. Oliver 2010)—in the emotional moment of a prestigious graduation, it becomes easy not to ask why none of the professors or the new PhDs use a wheelchair or a sign language interpreter.

come saddled with ableist stereotypes. STEM students who call attention to their diverse needs risk disrupting their under-construction discourse identities as precise, up-to-standard professionals (Slaton 2013).

The DisCrit and crip theory literature shows how applying the term "identity" to gender, race, disability, and scientific field often involves shifting among Gee's four connotations in terms of where the identity derives from and how it is acted out. In our study of disability in physics, we are especially drawn to the institutional and discourse senses. Specifically,

1. How have people's experiences as students or physicists with disabilities played out in the context of university or workplace power structures?
2. Have they interacted with a formal "accommodations" structure, and if so, what were the results?
3. How have their relationships with other people (peers, mentors, advisees, staff, etc.) been a part of their trajectory?

These two literatures explore issues of power and recognition, who defines "normal" and what that boundary can mean for those who fall variously outside it.

8.3.4 Disability Identity Reprise—Weaving Together the Threads

We began this section by marking a split between medical and social models of disability. Like all stark binaries, this idea unfolds into complexity. Shakespeare (1996) unpacks five variants of the social model, noting that the unified, unequivocal positions needed for political activism are often explored in more nuance by those on the "inside." He draws some parallels between the disability rights movement and the struggles for women's and racial civil rights (Linton 1998). However, Shakespeare's chapter also notes divergence: compared to gender or racial identity, an identity as a person with a disability may cause isolation in families. Someone with a disability may face ongoing decisions about whether to "come out" about their disability in various social settings. These experiences may be more parallel to those of scientists in the LGBT community, and indeed a "coming out" metaphor recurs in disability studies.

One final question which we argue that education researchers should return to is: What is the goal of this work? Some accessibility efforts are underpinned by the hope that in the future, no such accommodations will be needed. If medical science can someday "fix" all impairments, would this not be ideal? This thought, usually unexpressed but often present, goes back to the notions of a compulsory standard of "normal" raised in crip theory (McRuer 2006a). We raise it here to disagree explicitly: the goal is not a future where neurodiversity and diversity of bodies has been somehow smoothed away[7]. As readers of older science fiction or futurism can testify, the future is always different—weirder, more varied, and in many ways more wonderful—than we can predict from the current point in time. Physics is capable of adapting to a more equitable and diverse body—and bodies—of students and practitioners. If we believe it is a truly fundamental science, then finding and welcoming a wide range of talented individuals must be a perennial priority.

The culture of STEM today is one that strives for the standardization of reasoning and behavior (Nespor 1994; Slaton 2013). There is little acceptance of, or interest in, differences.[8] It is also true that people in STEM are seen as White, male, cis, heterosexual, and able-bodied. These perceptions can exclude anyone who might be different: Why go into a field where you are the only one of your identity? Those who do push for change are labelled as outsiders and often ostracized by others in their field (Slaton 2013).

Ability, like race, is a social construct. The experts are those in power, and their knowledge about not only STEM but about who is able to do STEM comes from them—so they remain in power. Any suggestion that the culture of STEM is not already fair is taken as a threat to American values of meritocracy and democracy (Slaton 2013).

People in STEM have traditionally published papers in the passive voice, not referring to themselves at all; therefore all identifiers of race, ability, gender, and sexuality are left out. This contributes to the lack of visibility in the field and upholds systems of oppression. Although many would argue that these identities do not influence the field and are therefore irrelevant, at the same time feminist theory, critical race theory, and DisCrit inform us that most people believe that women, people of color, and people with a disability are less capable in STEM. The medical and binary models of both gender and disability keep these systems in place (Slaton 2013).

8.4 Identity and Intersectionality

One of the challenges of categorizing identity is that imposing categories invites us to look at identity or oppression as a linear problem. If we can understand the state of "gender," and the state of "race," then the superposition of these two states tells us what is going on or how to fix the injustice. There's a high risk of doing this when only one category is considered. The phrase "gender in physics" implies a world where "gender" is a thing with a universal essence, experienced in more or less similar ways by all varieties of people. Even when considering gender and race at the same time, the linear superposition model is easy to apply, often without even

[7] For essays and fiction exploring this point, see Uncanny Magazine's *Disabled People Destroy Science Fiction!* issue (Sjunneson-Henry et al. 2018).

[8] At least, difference between researchers—even when the goal is a relentless search for differences among research subjects (C. N. Jacklin 1981).

noticing. Exploring the nonlinear interactions of lived experiences takes an explicitly different way of thinking about identity categories. This problem has been explored in various ways by feminist scholars using frameworks of intersectionality (McCall 2005). In this section, we want to focus on that intersectionality, which has already occurred by example in some of the frameworks discussed above. We begin with work in physics, then discuss how intersectionality is used in DisCrit and crip theory. Gender is a theme in some of these research areas, but not all, and we return to this point in Sect. 8.5.

8.4.1 Beginnings and the Double Bind

Studying intersectionality has a number of roots, one of them in scholarship about the blend of racism and sexism experienced by women of color (Crenshaw 1991). Crenshaw's work discusses how social programs designed to help women (with the implication of "all women") are often structured so that they are most useful to White women because of assumptions made by the (usually White and affluent) benefactors about language access, family childcare resources, and other factors. In STEM fields and in physics, Ong et al. (2011) and Ko et al. (2013) explore the idea of the "double bind," the simultaneous experience of racism and sexism that is more than simply an additive sum. Ko et al. (2013) examine themes of activism and work-life balance among women of color in physics and astrophysics. Resonating with Crenshaw's points, they note that work-life balance supports designed for White women in STEM may be less useful if women of color have different family roles or social pressures. In their narrative analysis, several women discussed taking less-prestigious career paths to be more available to mentor students, a choice they are more likely to face than White women academics (Prescod-Weinstein 2015).

Heidi Carlone and Angela Johnson studied fifteen women of color for several years, starting when they were undergraduates. All of these women majored in science, though none in physics. Although nearly all of them thought of themselves as scientists, most of them had interactions with other people in which they felt invisible, or worse, singled out in a negative way. Sometimes the students associated these negative experiences with their gender, sometimes with their race or ethnicity, and sometimes they said that many students had similar experiences (Carlone and Johnson 2007). Unfortunately, many of the themes that emerge from the studies in this section are about barriers, even when successful women are the focus of the research. In male-dominated fields, challenges to the "belonging-ness" of women are common. As gender combines with other facets of identity, these challenges often intensify for those positioned farther away from the "norm."

8.4.2 Gender and Race in Physics: Coping Methods, Shaping Methodologies

Several strands of work in physics have highlighted the intersection between gender and race. Maria Ong (2005) gives a longitudinal view of ten women of color in physics and the conflicts they encountered in reconciling field of study, gender, and race or ethnicity. Ong describes two broad strategies of fragmentation or multiplicity. In fragmentation, the women suppressed one or more facets of identity to "pass" as normal in a given context, such as buying pants on the way to lab to avoid comments from labmates, or adopting (though not necessarily endorsing) the kind of "zero uncertainty" language that is more associated with masculine forms of communication in the United States. In multiplicity, the women carved out ways to foreground their identities without suppressing them, such as a student who consciously adopted the role of "loud black girl" to claim space in meetings. Both of these strategies come with a price. Fragmentation means denying a part of yourself, which is a harsh cost to pay to practice science, while multiplicity tactics can provoke backlash from peers (as in the case of a woman who felt empowered to be more feminine at work, and promptly received pushback from her labmates despite her excellent work). There are echoes of this choice in talks given by disabled physicists at the 2016 American Association of Physics Teachers summer meeting[9] as well as in non-STEM specific research on disabled students (Jacklin 2011). Students who assert themselves to get accommodations for disabilities, regardless of on-paper legal protections, risk being stigmatized as "making excuses" or even being barred from programs. On the other hand, pursuing a fragmentation strategy of hiding impairments means that the student is at a hidden disadvantage when completing the same class and research tasks as their peers.

The work above uses intersectionality to focus the area of inquiry, with the goal of learning more about an understudied group of people. Other work embeds intersectionality even more deeply in the research methods. Rosa and Mensah (2016) use critical race theory (CRT) to study the pathways of Black women physicists. Their work highlights three themes from CRT: racism as a permanent feature in America, the importance of counter-storytelling against the dominant narrative, and interest convergence. The persistence of racism emerged in stories of their participants, who talked about exclusion from the graduate study groups formed by their White or Asian peers. This exclusion, a more pervasive theme than in studies focused only on gender, meant that it was sometimes easier to form a study group with other people of color or with international students outside of physics entirely. Counter-storytelling highlights these barriers but also the strategies that the women pursued to overcome them, and to counter dominant expectations ("[she] probably won't amount to anything," in the words of one teacher (p. 6)). Disabled scientists face their own range of dominant narratives about their ability to do their work, and Rosa and Mensah's work shows how researchers can choose

[9] Session BJ, http://www.aapt.org/Conferences/sm2016/session.cfm.

methods that elevate the stories of their participants over an oppressive system. Finally, interest convergence is the idea that civil rights advances for Black Americans only occur when they serve the interests of White Americans as well, as in the case of school desegregation (championed for years by African Americans) only happening in the face of Cold War labor and public image demands. There are echoes of this theme in the Universal Design framework, which argues that making classes more accessible for disabled students will benefit everyone in the long run (Burgstahler 2015; James et al. 2018). For example, posting lecture notes may help students with executive function issues to stay engaged in class, but also help other students who can use the example of a well-structured outline. While true and important, this argument must be made by Universal Design advocates because a justice-based argument is historically not enough.

Other work by Simone Hyater-Adams and collaborators (Hyater-Adams et al. 2018) fuses prior work on gender and physics identity with constructs of racialized identity to tease out complexities in students' accounts of their experiences. These authors give extensive detail about building their framework, providing an example of what this process can look like in physics. Problematizing research methods is one of the hardest ways to do feminist physics, and we are profoundly grateful to the authors in this section who are leading the way.

8.4.3 Race and Disability: Institutional Intersectionality

The emerging field of DisCrit uses the interplay between critical race theory and disability studies to examine how diagnoses and experiences of disability play out in racialized ways. In school, people of color are overrepresented in special education classes (Artiles et al. 2010; Tomlinson 2016). Connor et al. (2016) argue that this is due to White supremacy. People of color are often labelled as having a learning/intellectual disability or mental health/behavioral issues based on the subjective views of their mostly White teachers, who have been socialized to see non-White people, even their students, as violent and/or intellectually lacking. It is worth noting that people of color are less likely to be overrepresented in physical or sensory disabilities, suggesting racism is a key player in labelling disability (Connor et al. 2016). DisCrit does not only focus on an institutional view, but this perspective is important to add to the above work in physics. Intersectional work in PER has tended to focus on the experiences of individual students and their interactions with other students or faculty members. The university as an edifice may or may not enter into these stories, but is likely to come up for any person who has to navigate an official accommodations process. Disability has a long history as an object of research, where universities built for the elite used people with disabilities as fuel for scientific output while excluding them from membership (Dolmage 2017). Much DisCrit work to date focuses on primary or secondary education, but Dolmage's work on the history of academic ableism warns us that these themes continue through college.

8.4.4 Disability and Sexuality: The Enforcement of "Normal"

Disability status interacts with other identities in interesting ways as well. McRuer (2006b) talks about the intersectionality of queerness and disability in his book *Crip Theory.* Examples he highlights include gay men living with HIV and the Sharon Kowalski case, where the courts had to intervene to allow Kowalski to recuperate with her same-sex partner after she was severely injured in a car accident. In a more recent book, Hirsch (2018) talks about the intersectionality of gender and disability. She talks about how many women, especially young women, do not "come out" and disclose their illnesses or disabilities—and even when they do, they minimize them. She also discusses further intersectionality with queerness, as she notes that female partners are less likely to leave disabled women than male partners are, and with being a woman of color, which makes navigating life as a person with a serious illness or disability harder. Justine Egner (2016) reviews crip theory from the perspective of medical sociology, perhaps the closest current reference to how disability and sexuality intersect with issues in science. There, she defines crip theory as "a disability focused queer approach that is concerned with the relationship between the physical body, embodiment, and the self" (p. 161). This perspective contrasts sharply with the prevailing "culture of no culture" in physics (Traweek 1988), where bringing up bodily needs and realities is taboo and seen as a distraction from pure science. Enger ends with a review of empirical research on disability that applies queer or crip perspectives (2016, pp. 186–187) and suggestions for future research directions. These include two that seem especially salient to studying disability in physics: How do individuals who identify as both LGBT and disabled negotiate these (at times) socially contradicting identities? How does the importance placed on progressive and curative discourses shape and affect disabled people's personal narratives and experiences?

8.4.5 Approaches to Intersectionality

McCall (2005) distinguishes three broad approaches to the complexity of categories in intersectional feminist work. The first, anticategorical, draws from poststructuralism (such as work by Foucault referenced earlier) and attacks the very usefulness of researchers imposing a set of categories. The intercategorical approach seeks to understand relationships of inequality along multiple axes at the same time (e.g. gender, race or ethnicity, and education level), combining women's studies insights with quantitative methods. Between these two approaches is what McCall calls intracategorical complexity, which understands that identity categories are provisional and somewhat reductionist, but nonetheless have social reality. Studies in this framework often focus on a particular social group or position that has been previously under-studied, with the goal of elevating those voices. This reading of intersectionality resonates with our goals in exploring disability in physics.

8.5 Beyond the Binary View of Gender and Disability

8.5.1 Reflecting on Frameworks

In her book *Reflections on Gender and Science*, Evelyn Fox Keller (1985), discusses the association that science has with the "impersonal, rational, and general" (p. 7). This has allied well, to the benefit of many, with the association of men with "objectivity, reason, and mind"—and not very well, to the detriment of many, with the association of women with "subjectivity, feeling, and nature" (p. 7). In physics education research, we have been on the margins between scientific research and social science research. There have, historically, been few distinctions between these research paradigms; both take as a given that there is a clear division between researchers and their subjects and that the researcher knows things that their subjects do not. When the subjects of research are people, however, these distinctions should be questioned (M. Oliver 1992; Whitten 2012). At one extreme, traditional social science research has been called "the rape model of research" (Reinharz 1979). In her *Feminist Methods in Social Research*, Reinharz (1992) argues for the amplification of less-heard voices and attention to the intersection of identities. Whitten argues that research in physics, not just in physics education, should embrace categories of projects to make physics more inclusive and to set physics research in social/political context (Whitten 2012).

The theoretical frameworks discussed above each bring tools for thinking critically about disability in physics. Social models of disability focus on structural inequities rather than interrogating students for "what is wrong with you?" (Table II, below). However, physics has aligned itself strongly with the "objectivity, reason, and mind" half of the binary discussed by Keller. That stance is friendly to the medicalization of disability discussed in crip theory (Egner 2016), because it neatly partitions disability as a phenomenon to be studied in isolation from the surroundings using familiar reductionist approaches. In fighting this cultural inertia, we may draw on some of the same arguments in favor of diversity in science that have already been made for gender and race. These arguments include the idea that science is underserved if large sections of its talent pool are excluded (Tilghman 2003). This is true, but it is logic that critical race theory might recognize as interest convergence—that things must improve for Black students, or students with disabilities, because that will better serve a White, able-bodied majority. In using these rhetorical framings for scientist peers or funding agencies, we must not lose sight of justice-based goals, even if we voice them less to those audiences.

DisCrit and crip theory can teach us about the intersections of disability with other facets of identity. These may be institutional (in terms of how people are categorized or "accommodated"), or play out in how people are recognized or treated by peers. The lessons from this work are bittersweet: it teaches us to see new kinds of injustice, but also to celebrate the complexity of human lives that cannot be contained (Star 1991).

8.5.2 *Reflecting on Methods*

One thing that often happens in quantitative research, even (or especially) by those with good intentions, is that some subject identities disappear. In many quantitative studies in science and engineering, there are often so few students who are not White men that researchers avoid disaggregating their data—they do not want to compromise the anonymity of their subjects. This means that, even if results for women are reported separately, or results for non-White people are reported separately, it is incredibly unlikely that, for example, results for women of color will be reported separately from those of White women (Slaton and Pawley 2015). "If quantitative human sciences research (whether deploying large or small numbers of subjects) relies on the use of categories (delineating white, black, or brown subjects; healthy or ill subjects; subjects of particular genders; students of various achievement levels, etc.) as the basics of its systematic inquiry, then recent Queer Theory prompts us to question the social origins and functions of category making" (Slaton and Pawley 2015, p. 26.1564.6). The "queering" of anything, including research categories, involves embracing contradictions.

There is already at least one good example, though being published in the journal *Disability and Society*, it has been hard for physicists to find. Gibson (2012) interviewed university students with disabilities, producing narrative accounts that illustrate the students' experiences of exclusion and of barriers to learning. This work aligns with the various calls above to prioritize the voices of students and others on the margins (Annamma et al. 2016; Reinharz and Davidman 1992; Rosa and Mensah 2016). Further, Oliver (1992) calls for us to be careful in how we ask questions of our subjects. Table 8.2 contrasts questions actually asked by the British government in a disability survey with possible alternatives.

These alternative questions, centering on the validity of the experiences of the people with disabilities, seem more likely to not only treat them with respect and dignity but describe those experiences so that others have a chance to understand

Table 8.2 Taken from Tables I and II in Oliver 1992

Sample of questions from a 1986 survey of disabled adults by the British government	Sample of alternative questions written by Oliver (1992)
1. Can you tell me what is wrong with you?	1. Can you tell me what is wrong with society?
2. Are your difficulties in understanding people mainly due to a hearing problem?	2. Are your difficulties in understanding people mainly due to their inabilities to communicate with you?
3. Have you attended a special school because of a long-term health problem or disability?	3. Have you attended a special school because of your education authority's policy of sending people with your health problem or disability to such places?
4. Does your health problem / disability affect your work in any way at present?	4. Do you have problems at work because of the physical environment or the attitudes of others?

them. We know from standpoint theory that there are some things people with privilege may never truly be able to understand (Hartsock 2003). Hartsock argues that the nature of women's work (reproduction and beyond) gives them a standpoint from which they can see things that men never could. The argument could certainly extend to other people who have traditionally been discriminated against: people of color, LGBTQ people, and people with disabilities. People who have not experienced discrimination do not, and perhaps cannot, perceive the same world. That being said, we hold out hope that qualitative research that respects those who have been excluded from physics can shine light on their experiences so that others have a chance to see (or hear, or feel).

Finally, though it may be pushing uphill, research questions can be framed to interrogate the system rather than the students. Scanlon and collaborators examine four research-based introductory physics curricula for alignment with the guidelines of Universal Design for Learning (Scanlon et al. 2018). They find that some checkpoints are well satisfied, but others (such as providing multiple means of engagement) are not. They give detailed suggestions for thinking about accessibility and Universal Design in curriculum development, both critiquing the physics education community and providing ways forward. Though there is much to be angry or discouraged about, we (the authors of this chapter) have trouble imagining this conversation happening 10 or 20 years ago. Change is necessary, but it is also possible.

8.5.3 Returning to Gender

In the literature discussed above, gender occurs in several places, most often in work on women of color. But gender has not been a central theme of our review, in part because searching for literature on "women in _____" tends to yield results that focus on the experiences of white, straight, cisgender women. Our goal was to explore scholarship surrounding an understudied-in-science identity, that of disability. When photographing the corona of the Sun, you need a filter (or an interposed moon) to block out the overwhelming brightness at the center. Here, to align different theoretical slants on disability and identity, we have first looked to the side of gender.

In the realm of lived experience, gender infuses and connects these other facets. Literature on feminist disability studies (Garland-Thomson 2005) explores these connections and emphasizes elevating unheard voices, challenging social constructions of disability and "normal," and drawing analogies between ableism and systems of racial or gender oppression. Much of this work lies in women's studies, literary criticism, or other scholarship that may not be read by physics education researchers. Some of the lessons translate bluntly: "Women with disabilities, even more intensely than women in general, have been cast in the collective cultural imagination as inferior, lacking, excessive, incapable, unfit, and useless" (Garland-Thomson 2005, p. 1567). For a career in science, you must make an identity—in the minds of others and yourself—that is the opposite of these qualities. Echoing Ong's finding of fragmentation as a self-preservation strategy, it is no surprise that many women in the sciences must hide, minimize, or deny disability as a piece of their identity. The effect of gender is not limited to women; disability is often seen as weakening men and making them less masculine. For agender or nonbinary scientists, disability can become another axis of their existence that must be translated and negotiated to deal with teachers or colleagues. The specifics vary from person to person; the only relative constant is that gender shapes how society reacts to disability. Though we have prioritized other writings in this chapter, gender is always there behind the filter.

8.5.4 Final Thoughts

It is hard to be anything but White, male, cisgender, straight, and able-bodied and be recognized as a physicist, and that is a shame. We already know that it is hard to embrace duel identities of women and physicist, and above we have tried to chart some of the space of disability identity. Students have to worry about whether to disclose their disabilities, the inaccessibility of laboratory and other work spaces, and the very slim chance that they will find a mentor who is "out" about being disabled. To think about how scientists at all career stages grapple with these issues, we have drawn together elements of disability studies, critical race theory, DisCrit, and crip theory. A depressingly common thread in these studies is that people across these identity groups are viewed as less capable in science. Though the categories of identity are socially constructed, they have real consequences for the people sorted into them (Lewontin et al. 1993; Mostert 2002). To break free of these old patterns, we must wield several key ideas:

8.5.4.1 The Interlinking of Disability and Other Facets of Identity

In classification, recognition, accommodation, or decisions not to disclose, gender and race and LGBT identity inevitably affect the institutions and people around you and how they react.

8.5.4.2 The Question of Who Gets to Belong and Who is "Normal"

Physics tends to enshrine certain kinds of genius—White, male, socially awkward—and not to look far outside this mold for talent. Challenging these ideas can provoke a backlash from physicists who came up through this tradition and now see attacks on their own identities as scientists.

8.5.4.3 The Importance of Voice

If everyone is different and even the concept of "normal" is weighed down with baggage, where should a researcher begin? The importance of being recognized, of telling your own story about yourself, comes up over and over. If we were to frame a narrative study of disability in physics at the size of a cross stitch sampler, it might be: "What is your story? Tell us about you and physics."

References

Adjunct, A. K. (2008). The revolving ramp: Disability and the new adjunct economy. *Disability Studies Quarterly, 28*(3). Retrieved from http://dsq-sds.org/article/view/110.

Americans with Disabilities Act Title II Regulations. (n.d.). Retrieved from https://www.ada.gov/regs2010/titleII_2010/titleII_2010_regulations.htm

Annamma, S. A., Connor, D. J., & Ferri, B. A. (2016). Dis/ability critical race studies (DisCrit): Theorizing at the intersections of race and dis/ability. In D. J. Connor, B. A. Ferri, & S. A. Annamma (Eds.), *DisCrit: Disability studies and critical race theory in education* (pp. 9–32). Teachers College Press.

Artiles, A. J., Kozleski, E. B., Trent, S. C., Osher, D., & Ortiz, A. (2010). Justifying and explaining disproportionality, 1968–2008: A critique of underlying views of culture. *Exceptional Children, 76*(3), 279–299. https://doi.org/10.1177/001440291007600303.

Atherton, T. J., Barthelemy, R. S., Deconinck, W., Falk, M. L., Garmon, S., Long, E., et al. (2016). *LGBT climate in physics: Building an inclusive community*. College Park, MD: American Physical Society.

Barad, K. (2007). *Meeting the universe halfway: Quantum physics and the entanglement of matter and meaning*. Duke University Press.

Berkeley Graduate and Professional Schools Mental Health Task Force. (2004). *Berkeley graduate student mental health survey*. Retrieved from University of California website: http://regents.universityofcalifornia.edu/regmeet/sept06/303attach.pdf

Brahmia, S. W. (2008). Improving learning for underrepresented groups in physics for engineering majors. *AIP Conference Proceedings, 1064*, 7–10. https://doi.org/10.1063/1.3021279.

Brickhouse, N. W., & Potter, J. T. (2001). Young women's scientific identity formation in an urban context. *Journal of Research in Science Teaching, 38*(8), 965–980. https://doi.org/10.1002/tea.1041.

Burgstahler, S. E. (Ed.). (2015). *Universal design in higher education: From principles to practice* (2nd ed.). Cambridge, MA: Harvard Education Press.

Carabajal, I. G., Marshall, A. M., & Atchison, C. L. (2017). A synthesis of instructional strategies in geoscience education literature that address barriers to inclusion for students with disabilities. *Journal of Geoscience Education, 65*(4), 531–541. https://doi.org/10.5408/16-211.1.

Carlone, H. B., & Johnson, A. (2007). Understanding the science experiences of successful women of color: Science identity as an analytic lens. *Journal of Research in Science Teaching, 44*(8), 1187–1218. https://doi.org/10.1002/tea.20237.

Committee on Equal Opportunities in Science and Engineering. (2009). *Broadening Participation in America's STEM Workforce: 2007–2008 CEOSE Biennial Report to Congress* (No. CEOSE 09-01). Retrieved from https://www.nsf.gov/od/oia/activities/ceose/index.jsp

Connor, D. J., Ferri, B. A., & Annamma, S. A. (2016). *DisCrit: Disability studies and critical race theory in education*. New York: Teachers College Press.

Crenshaw, K. (1991). Mapping the margins: Intersectionality, identity politics, and violence against women of color. *Stanford Law Review, 43*(6), 1241–1299. https://doi.org/10.2307/1229039.

Crenshaw, K., Gotanda, N., Peller, G., & Thomas, K. (Eds.). (1995). *Critical race theory: The key writings that formed the movement*. New York: The New Press.

Danielsson, A. T. (2010). Gender in physics education research: A review and a look forward. In M. Blomqvist & E. Ehnsmyr (Eds.), *Never mind the gap! Gendering science in transgressive encounters* pp. 65–83. University Printers, Uppsala.

Daston, L. (1992). Objectivity and the escape from perspective. *Social Studies of Science, 22*(4), 597–618. https://doi.org/10.1177/030631292022004002.

Delgado, R. (1995). *Critical race theory: The cutting edge*. Philadelphia: Temple University Press.

Delgado, R., & Stefancic, J. (2012). *Critical race theory: An introduction* (2nd ed.). New York: NYU Press.

Dolmage, J. T. (2017). *Academic Ableism*. Ann Arbor, MI: University of Michigan Press.

Egner, J. (2016). A messy trajectory: From medical sociology to crip theory. In B. Altman & S. Barnartt (Series Ed.), *Research in social science and disability: Vol. 9. Sociology looking at disability:* (pp. 159–192). https://doi.org/10.1108/S1479-354720160000009009

Essed, P. (1990). *Everyday racism: Reports from women of two cultures*. Claremont, CA: Hunter House.

Faulkner, W. (2009). Doing gender in engineering workplace cultures. II. Gender in/authenticity and the in/visibility paradox. *Engineering Studies, 1*(3), 169–189. https://doi.org/10.1080/19378620903225059.

Foucault, M. (1979). *Discipline and punish: The birth of the prison*. New York: Vintage Books.

Garland-Thomson, R. (2005). Feminist disability studies. *Signs: Journal of Women in Culture and Society, 30*(2), 1557–1587. https://doi.org/10.1086/423352.

Gee, J. P. (2000). Identity as an analytic lens for research in education. *Review of Research in Education, 25*, 99–125. https://doi.org/10.2307/1167322.

Gibson, S. (2012). Narrative accounts of university education: Socio-cultural perspectives of students with disabilities. *Disability & Society, 27*(3), 353–369. https://doi.org/10.1080/09687599.2012.654987.

Gieryn, T. F. (1983). Boundary-work and the demarcation of science from non-science: Strains and interests in professional ideologies of scientists. *American Sociological Review, 48*(6), 781–795. https://doi.org/10.2307/2095325.

Gonsalves, A. J. (2014). "Physics and the girly girl—There is a contradiction somewhere": Doctoral students' positioning around discourses of gender and competence in physics. *Cultural Studies of Science Education, 9*, 503–521. https://doi.org/10.1007/s11422-012-9447-6.

Gould, S. J. (1996). *The mismeasure of man*. New York: W. W. Norton & Company.

Harding, S. (1986). *The Science Question in Feminism*. Ithaca, NY: Cornell University Press.

Hartsock, N. C. (2003). In S. Harding & M. B. Hintikka (Eds.), *The feminist standpoint: Developing the ground for a specifically feminist historical materialism*. Dordrecht: Kluwer Academic Publishers.

Hill, C., Corbett, C., & St. Rose, A. (2010). *Why so few? Women in science, technology, engineering, and mathematics*. Retrieved from http://www.aauw.org/research/why-so-few/

Hirsch, M. L. (2018). *Invisible: How young women with serious health issues navigate work, relationships, and the pressure to seem just fine*. Boston, MA: Beacon Press.

Hyater-Adams, S., Fracchiolla, C., Finkelstein, N., & Hinko, K. (2018). Critical look at physics identity: An operationalized framework for examining race and physics identity. *Physical Review Physics Education Research, 14*(1), 010132. https://doi.org/10.1103/PhysRevPhysEducRes.14.010132.

Jacklin, A. (2011). To be or not to be 'a disabled student' in higher education: The case of a postgraduate 'non-declaring' (disabled) student. *Journal of Research in Special Educational Needs, 11*(2), 99–106. https://doi.org/10.1111/j.1471-3802.2010.01157.x.

Jacklin, C. N. (1981). Methodological issues in the study of sex-related differences. *Developmental Review, 1*(3), 266–273. https://doi.org/10.1016/0273-2297(81)90021-6.

James, W., Lamons, K., Schreffler, J., Vasquez, E., III, & Chini, J. J. (2018). Exploring learner variability: Experiences of students with cognitive disabilities in postsecondary STEM. In L. Ding, A. Traxler, & Y. Cao (Eds.), 2017 *Physics Education Research Conference proceedings* (pp. 192–195). https://doi.org/10.1119/perc.2017.pr.043.

Kanim, S., & Cid, X. C. (2017). The demographics of physics education research. *ArXiv:1710.02598 [Physics]*. Retrieved from http://arxiv.org/abs/1710.02598

Keller, E. F. (1985). *Reflections on gender and science*. New Haven: Yale University Press.

Ko, L. T., Kachchaf, R. R., Ong, M., & Hodari, A. K. (2013). Narratives of the double bind: Intersectionality in life stories of women of color in physics, astrophysics and astronomy. *AIP Conference Proceedings, 1513*, 222–225. https://doi.org/10.1063/1.4789692.

Ladson-Billings, G. (1998). Just what is critical race theory, and what's it doing in a nice field like education? *International Journal of Qualitative Studies in Education, 11*(1), 7–24. https://doi.org/10.1080/095183998236863.

Lee, A. (2014). Students with disabilities choosing science technology engineering and math (STEM) majors in postsecondary institutions. *Journal of Postsecondary Education and Disability, 27*(3), 261–272.

Levecque, K., Anseel, F., De Beuckelaer, A., Van der Heyden, J., & Gisle, L. (2017). Work organization and mental health problems in PhD students. *Research Policy, 46*(4), 868–879. https://doi.org/10.1016/j.respol.2017.02.008.

Lewontin, R. C., Rose, S., & Kamin, L. J. (1993). IQ: The rank ordering of the world. In S. G. Harding (Ed.), *The "racial" economy of science: Toward a democratic future*. Bloomington: Indiana University Press.

Liasidou, A. (2014). Critical disability studies and socially just change in higher education. *British Journal of Special Education, 41*(2), 120–135. https://doi.org/10.1111/1467-8578.12063.

Linton, S. (1998). Disability studies/not disability studies. *Disability & Society, 13*(4), 525–539. https://doi.org/10.1080/09687599826588.

Marks, G. S. (2017, December 15). *It's time to stop excluding people with disabilities from science*. Retrieved September 21, 2018, from Massive website: https://massivesci.com/articles/disability-science-career-stem-field

McCall, L. (2005). The complexity of intersectionality. *Signs: Journal of Women in Culture and Society, 30*(3), 1771–1800. https://doi.org/10.1086/426800.

McRuer, R. (2006a). Compulsory able-bodiedness and queer/disabled existence. In L. J. Davis (Ed.), *The disability studies reader (Second)*. New York: Taylor & Francis.

McRuer, R. (2006b). Crip theory: Cultural signs of queerness and disability. New York: NYU Press.

Mostert, M. P. (2002). Useless eaters: Disability as genocidal marker in Nazi Germany. *The Journal of Special Education, 36*(3), 157–170. https://doi.org/10.1177/00224669020360030601.

National Science Foundation, & National Center for Science and Engineering Statistics. (2017). *Women, minorities, and persons with disabilities in science and engineering: 2017* (No. NSF 17-310). Retrieved from www.nsf.gov/statistics/wmpd/

Nespor, J. (1994). *Knowledge in motion: Space, time, and curriculum in undergraduate physics and management*. London: Falmer Press.

Oliver, M. (1992). Changing the social relations of research production? *Disability, Handicap & Society, 7*(2), 101–114. https://doi.org/10.1080/02674649266780141.

Oliver, P. (2010). *Foucault: The key ideas*. London: Hodder Education.

Ong, M. (2005). Body projects of young women of color in physics: Intersections of gender, race, and science. *Social Problems, 52*(4), 593–617. https://doi.org/10.1525/sp.2005.52.4.593.

Ong, M., Wright, C., Espinosa, L. L., & Orfield, G. (2011). Inside the double bind: A synthesis of empirical research on undergraduate and graduate women of color in science, technology, engineering, and mathematics. *Harvard Educational Review, 81*(2), 172–209. https://doi.org/10.17763/haer.81.2.t022245n7x4752v2

Paul, S. (2000). Students with disabilities in higher education: A review of the literature. *College Student Journal, 34*(2), 200.

Prescod-Weinstein, C. (2015, November 18). *Still working for free: The Unpaid labor of blackademics*. Retrieved April 30, 2018, from Medium website: https://medium.com/thsppl/still-working-for-free-the-unpaid-labor-of-blackademics-dfa84c200ff0

Price, M., Salzer, M. S., O'Shea, A., & Kerschbaum, S. L. (2017). Disclosure of mental disability by college and university faculty: The negotiation of accommodations, supports, and barriers. *Disability Studies Quarterly, 37*(2). https://doi.org/10.18061/dsq.v37i2.5487.

Reinharz, S. (1979). *On becoming a social scientist.* San Francisco: Jossey-Bass.

Reinharz, S., & Davidman, L. (1992). *Feminist methods in social research.* New York: Oxford University Press.

Rosa, K., & Mensah, F. M. (2016). Educational pathways of Black women physicists: Stories of experiencing and overcoming obstacles in life. *Physical Review Physics Education Research, 12*(2), 020113. https://doi.org/10.1103/PhysRevPhysEducRes.12.020113.

Scanlon, E., Schreffler, J., James, W., Vasquez, E., & Chini, J. J. (2018). Postsecondary physics curricula and Universal Design for Learning: Planning for diverse learners. *Physical Review Physics Education Research, 14*(2), 020101. https://doi.org/10.1103/PhysRevPhysEducRes.14.020101.

Sevo, R. (2012). Recommended reading: Disabilities and diversity in science and engineering. In B. Bogue & E. Cady (Eds.), *Apply research to practice (ARP) resources*. Retrieved from http://www.engr.psu.edu/AWE/ARPResources.aspx.

Shakespeare, T. (1996). Disability, identity, and difference. In C. Barnes & G. Mercer (Eds.), *Exploring the divide* (pp. 94–113). Leeds: The Disability Press.

Sjunneson-Henry, E., Parisien, D., Barischoff, N., Qiouyi Lu, S., & Tarr, J. (Eds.). (2018). Disabled people destroy science fiction! *Uncanny Magazine*, (24). Retrieved from https://uncannymagazine.com/issues/uncanny-magazine-issue-twenty-four/

Slaton, A. E. (2013). Body? What body? Considering ability and disability in STEM disciplines. *2013 ASEE Annual Conference & Exposition*, 23.247.1–23.247.16. Retrieved from https://peer.asee.org/19261.

Slaton, A. E., & Pawley, A. L. (2015). The power and politics of STEM research design: Saving the "small N." *2015 ASEE Annual Conference & Exposition*, 26.1564.1–26.1564.13. Retrieved from https://peer.asee.org/24901

Star, S. (1991). Power, technology and the phenomenology of conventions: On being allergic to onions. In J. Law (Ed.), *A sociology of monsters: Essays on power, technology, and domination* (pp. 26–56). Routledge.

Sutton, H. (2017). Students with disabilities as likely to enter STEM fields as those without disabilities. *Disability Compliance for Higher Education, 22*(9), 9–9. https://doi.org/10.1002/dhe.30292.

The Graduate Assembly. (2015). *Graduate student happiness & well-being report 2014*. Retrieved from University of California website: http://ga.berkeley.edu/wellbeingreport/

Tilghman, S. M. (2003, October). *The challenges of educating the next generation of the professoriate*. The 2003 Killam Lecture presented at the University of British Columbia. Retrieved from https://www.princeton.edu/president/tilghman/speeches/20031023/

Tomlinson, S. (2016). Race, class, ability, and school reform. In D. J. Connor, B. A. Ferri, & S. A. Annamma (Eds.), *DisCrit: Disability studies and critical race theory in education* (pp. 157–166). New York: Teachers College Press.

Traweek, S. (1988). *Beamtimes and lifetimes: The world of high energy physicists*. Cambridge, MA: Harvard University Press.

Traxler, A. L., Cid, X. C., Blue, J., & Barthelemy, R. (2016). Enriching gender in physics education research: A binary past and a complex future. *Physical Review Physics Education Research, 12*, 020114. https://doi.org/10.1103/PhysRevPhysEducRes.12.020114.

United States Department of Labor. (n.d.). *Voluntary self–identification of disability form*. Retrieved September 29, 2018, from https://www.dol.gov/ofccp/regs/compliance/sec503/self_id_forms/selfidforms.htm

Whitten, B. L. (1996). What physics is fundamental physics? Feminist implications of physicists' debate over the superconducting supercollider. *NWSA Journal, 8*(2), 1–16.

Whitten, B. L. (2012). (Baby) Steps toward feminist physics. *Journal of Women and Minorities in Science and Engineering, 18*(2), 115–134. https://doi.org/10.1615/JWomenMinorScienEng.2012003648.

Wilcox, C. (2014, October 10). *Lighting dark: Fixing academia's mental health problem*. Retrieved September 29, 2018, from New Scientist website: https://www.newscientist.com/article/dn26365-lighting-dark-fixing-academias-mental-health-problem/

Wilkie, D. (2014, August 7). *Challenges confront disabled who pursue STEM careers*. Retrieved September 21, 2018, from SHRM website: https://www.shrm.org/resourcesandtools/hr-topics/behavioral-competencies/global-and-cultural-effectiveness/pages/disabled-in-stem-careers.aspx

Wright State University Office of Disability Services. (2018). *Online student application*. Retrieved September 29, 2018, from https://denali.accessiblelearning.com/Wright/ApplicationStudent.aspx

Chapter 9
Elite British Female Physicists: Social Mobility and Identity Negotiations

Jaimie Miller-Friedmann

9.1 Introduction

Whilst recent years have seen improvement in women's participation in the broad field of science, there is still work to be done to achieve gender equity in all scientific fields. This is particularly the case in physics; in the UK, women remain the minority from undergraduate to professorship, despite small statistical increases in undergraduate participation between 1970 and the present. Further, the majority of those who participate in science self-identify as middle class. The literature cites many reasons why women and working-class students do not choose to participate in physics, and why they may leave the field. By contrast, there has been little research investigating why women and/or working-class women choose to stay in physics or how they have negotiated obstacles to achieve external esteem and success. This chapter reports on qualitative data from a study exploring the strategies and tactics used by female academics who have had significant success in the field of physics. The researcher used semi-structured, guided life-history interviews to examine how the six women in the sample negotiated obstacles to become international successes in their field. Analysis of the data revealed three significant experiences and identity negotiations that helped the women in this cohort persist in their determination to become academic physicists and to garner accolades and awards: (1) reliance on the self, (2) social support networks, and (3) the construction of a working class hero identity. The implications for the recruitment and retention of UK women in physics are discussed, along with suggestions for policy and practice.

J. Miller-Friedmann (✉)
University of Oxford, Oxford, UK
e-mail: J.L.Miller-Friedmann@bham.ac.uk

A. J. Gonsalves, A. T. Danielsson (eds.), *Physics Education and Gender*,
Cultural Studies of Science Education 19,
https://doi.org/10.1007/978-3-030-41933-2_9

9.2 Approaches to the Problem

Despite efforts in recent years to address the lack of gender equity in physics, both the numbers and the proportions of women's participation in the UK remain low. Existing research focuses on why women choose not to participate in physics (see Baram-Tsabari and Yarden 2008; Götschel 2013; Kantaria 2012; Stewart 1998), addresses women's participation in science as a general discipline (see Carli et al. 2016; Sikora and Pokropek 2012; Venville et al. 2013), or uses quantitative methods to correlate or predict women's participation (see Hazari et al. 2007; Lock et al. 2013; Nosek et al. 2009; Sadler et al. 2012). This work has been useful for identifying physics as a field rife with socially and academically preconceived notions that favour men's over women's participation. This occurs in a variety of ways—from discriminatory behaviours, to unconscious bias resulting in decreases in self-efficacy or interest. Among other reasons, many women appear to leave physics because they feel that they don't "belong"; their alienation from other (male) physicists makes them feel as though they will always be an outsider to the field.

To investigate this concept of belonging, scholars have analysed and (re)defined the notion of a *physics identity*. From an examination of career aspirations in young children, it is clear that those who tend to persist in physics self-identify as a physicist or as a "physics person" (Archer et al. 2016; Danielsson 2012; Gonsalves 2014; Gonsalves et al. 2013; Johnson et al. 2011). Constructing and adopting this identity seems to help individuals "fit in". Yet a physics identity is often in conflict with a feminine gender identity, and for this there seems to be no easy resolution. The conflict between identities may explain the high attrition of, and low recruitment for, women in physics (Archer et al. 2012; Danielsson 2012). Women often feel as though they must negotiate and potentially compromise their femininity to reconcile feminine and physics identities (Carli et al. 2016; Hazari et al. 2007; Lock et al. 2012). The struggle to achieve a unified self can be overwhelming and laborious, and may be interpreted as a reason to leave the field.

Recent studies have also investigated the intersection between class identity and aspirations towards a physics career. The UK 2007 Department for Education and Skills (DfES) report demonstrated that social class accounts for a great deal of variance in achievement and aspiration, and that working-class males tend to perform (and celebrate) a masculinity that is in direct conflict with academic success (DfES 2007). The ASPIRES project found that working-class students were not recognized as being naturally clever; cleverness was associated with white middle-class males, and because science tends to be highly correlated with cleverness, the field seems to "fit" white middle-class males (Archer et al. 2013; DeWitt et al. 2014). Working-class masculinity (perceived then, to be at odds with a science identity) is exemplified by the concept of *laddishness*, which is characterized by: performances of disruptive behaviour; the objectification of women through banter and "having a laugh"; and in men's appearance or avoidance of exerting efforts towards education (Carlone et al. 2015; Miller-Friedmann et al. 2018; Phipps and Young 2015; Stentiford 2018). Working-class students are further dissuaded from pursuing science by teachers' stereotypes and the inherent inequality in the UK school system (Archer et al. 2013). Because working-class women associate femininity with being a "mother" and "homemaker", they are even less likely than working-class boys to participate in science. A further disadvantage is that the higher education required for a career in science clashes with these ideal feminine identities (Fuller 2018).

There has been far less attention paid in the literature to women who have achieved success in academic physics, and to the ways in which they have managed real or perceived obstacles to their progress. Looking at gender inequality from this point of view is challenging because there are a limited number of women who have achieved internationally-acclaimed success in physics. In addition, researching this elite group of women would require anonymity on their part to ensure their candour does not threaten their external esteem and position. One notable study investigated scientists in elite universities and found that male and female scientists give different reasons for why gender inequality persists. The research concluded that gendered identity functions as a master narrative (over a science identity), and that women seem to be drawn to the kind of science that resonates with emotional labour (Ecklund et al. 2012). There are also few studies that examine the intersection between working-class women and academic physics, or explore the ways that women in this field negotiate, (re)mould, compromise, and leverage identities to construct a unified self that can achieve professional success.

Using Judith Butler's (1990) theorisation of gender, it is possible to investigate pathways to success for elite female physicists, and better understand the ways in which they have managed the challenges presented in the literature as reasons for attrition. According to Butler (1990), femininity is not a unified concept, but rather a contextual construct, (re)created and sustained through repeated acts and performances. Through the repetition of performance, an individual eventually adopts the identity that the performance implies, as well as the subjectivity that accompanies that identity (Butler 1990; Butler 1993). Butler (1990) acknowledges that notions of gender, and therefore performance and identity are contextual, although she does seem to indicate that there is an overarching master narrative of both femininity and masculinity. Becky Francis (2012), on the other hand, describes boys (males) as performing multiple competing masculinities, and girls (females) performing multiple competing femininities. The mixing and fusing of a variety of dichotomously associated gender performances is called gender *heteroglossia*—when one exists somewhere along the spectrum between masculine and feminine (Francis 2012). Gendered identity in this study will be discussed using dichotomous (monoglossia) terms, masculine and feminine, whilst acknowledging that the participants in the study, as is the case with most people in general, may experience gender heteroglossia. To be sure, most individuals in a society remain attached to gender dichotomy and tend to identify as either male or female, and the participants in this study self-identified as female. Their narratives, however, revealed a different gendered identity that provided insight into their professional success. Brickhouse et al.' (2000) description of identity "accounts for the importance of both individual agency as well as societal structures that constrain individual possibilities," thus positioning identity as both an internal ("Who am I?") and external ("How can that person be identified?") pursuit (Brickhouse et al. 2000). Acknowledging both the internal compulsion to fit in and external influences to "be normal", this study will analyse common experiences and coping mechanisms through the lens of gendered identity construction, as these identities are formed through imitation, repetition of performance and adoption.

Deterrents to women's participation in physics, such as gender stereotypes and lack of support in the workplace, have been recognised for a long time and are well researched. Yet actions to counteract these constraints have not resolved the problem of persistent gender inequality. By exploring the experiences of highly successful UK female physicists, this work looks at what constrains women from taking part in physics and, conversely, what has helped them to participate. Their narratives reveal previously ignored factors that have helped women in the UK stay the course in a physics department. A closer examination of their experiences also reveals possible methods for negotiating obstructions or difficulties.

9.3 The Study

To establish the current state of gender inequality in physics in the UK, and to determine which factors would constitute success, I investigated the number of women participating in physics in the UK. Females comprise 23% of physics undergraduate and postgraduate students, 24% of junior faculty, 17.4% of senior faculty, and 12.5% of Professors for the 2016/17 academic year (Higher Education Statistics Agency 2016). These statistics do indeed show that there is gender inequality in UK academic physics, but questions remain: Who are the women in the 12.5%, and how have they managed to circumvent obstacles? The literature gives abundant reasons why women would leave physics, if they decide to participate at all. What can be learned from the small proportion of women who have persisted?

9.4 Research Questions

To begin investigating what experiences senior women in academic physics in the UK have had, and how these experiences have influenced them and their professional pathways, the following questions were explored:

1. For those British women who have persisted through beginning stages and become academic physicists, in which ways have they experienced gender inequality, and which experiences have helped them succeed?
2. What effect has social class had on the participants, and has it been a hindrance or advantage in their pathway towards success?

9.5 Research Methods

9.5.1 Selection and Participants

The research consisted of six qualitative case studies of highly successful, female British academic physicists to investigate the experiences, coping mechanisms, and identity negotiations that have helped them navigate to their current elite status. This study defined successful and elite as those with "close proximity to power or particular professional expertise" (Lancaster 2016, p. 93) who have also achieved international and measurable success in physics. The participants for the study were chosen using the following parameters:

1. Applied for or been awarded the title of Professor (in the UK, 'Professor' is the highest academic rank, and a title of distinction awarded to an individual)
2. Been awarded a competitive grant from an internationally recognised organisation and/or scheme (such as the European Research Council [ERC] grants), *or*
3. Been recognised for her scientific achievements by an organisation or institution outside of her university (such as the Royal Society, or Women in Science and Engineering [WISE]) (Miller-Friedmann et al. 2018)

The number of women who have achieved any of these parameters in physics in the UK is small; the participants in this study met at least two and often all three. The participants can, therefore, be referred to as elite (as defined by this study), and successful (as they would likely be defined by their subfield peers).

The participants were selected by cross-referencing the Royal Society Fellowship directory and the European Research Council grant directory. The list of possible participants was narrowed down to those fitting at least two parameters, resulting in a possible pool of 12 candidates, who were invited to be interviewed. Six of the participants responded that they were interested in being interviewed, and arrangements were made to meet. The remaining six either declined the invitation to participate or never responded to attempts to contact. The six participants in this study were all British female physicists who were, or recently retired from being, faculty at internationally competitive research universities. All the participants were White, heterosexual, and five of the six participants had children. The participants ranged in age from 35 to 75; two participants were in the early to mid-career stage, and four were in their mid to late career stage.

9.5.2 Interviews and Analysis

The study was designed to be both confirmatory and exploratory— substantiating the obstacles reported in the literature, *and* discovering how the participants managed these obstacles. Qualitative interviews were chosen as a method of investigation for several reasons. First, the interview is a robust method for exploring meaning and explanation behind the numbers (Creswell 2003; Creswell and Creswell 2013; Gill et al. 2008). Second, in an effort to discover why women would stay rather than leave, I hoped to both uncover new data and give agency and voice to a minority population. The interview is an appropriate method to enable one to "hear silenced voices" (Creswell and Creswell 2013, page ref). The final reason for using qualitative interviews was to better understand the experiences of women, a critical perspective, as I was approaching the study from a feminist standpoint. Feminist researchers argue that women's life history narratives as constructed by women provide better insight into "women's lives, men's lives, and the whole social order" (Harding 1993, p. 56; see also Hughes 2001; Sinnes 2006). In this research, I used guided interviews to ask specific research questions, and because I sought to compare experiences. I guided the participants through semi-structured life history interviews, looking at the depth and breadth of their experiences, their coping mechanisms, and identity negotiations. The interview schedule was constructed to reflect both confirmation and exploration, and included questions that confirmed theories regarding self-efficacy, support, discrimination, and bias, as well as items that investigated possible reasons for success, such as attitude, gender associations, and positive mentoring. The participants were asked to block out 2 h for the interview. In reality the duration of the interviews varied from 2 h to close to three and a half hours. All interviews were followed with an online follow-up survey on surveymonkey.com. This anonymous survey consisted of six (essay) thought items, including: How do you define success?; Do you feel like gender stereotypes are different now to when you were a child?; and, one Likert scale item. The interviews were transcribed and uploaded into NVivo for analysis. Provisional codes (Creswell and Creswell 2013; Dey 1993) were used deductively, and inductive codes were used as they emerged in the data. All the transcripts were read, discussed, and coded by multiple researchers in order to establish reliability. The data were triangulated with the open-ended questions from the follow-up survey, published literature, and discussion with identity scholars in the science education community. All data were determined to be reliable and valid. The names used in this chapter are pseudonyms.

9.6 Findings: Identity Negotiations and Social Mobility

Analysis of the data demonstrated that there were three significant factors common to the participants that helped them participate and persist in academic physics and shift their social class. First, the participants experienced a prolonged period of isolation in their childhoods that impelled them to construct the beginnings of a non-dichotomous gendered identity, the characteristics of which were crucial to their later success. Second, the participants were highly involved in religious and musical groups that provided alternative ideologies to the norm they knew in school, and the alternative philosophies they performed were equally critical to both constructing their identities and succeeding in their careers. Last, the participants' narratives involved the trope of working-class hero, and this was a fundamental element to their identities, their perception of self, and their understandings of success.

9.6.1 From Isolation to Reliance on the Self

The participants began their life histories with descriptions of feeling isolated or excluded from their perception of the norm, and detailed the ways in which they coped with their difference. These experiences and subsequent coping strategies appeared to be critically important in guiding the participants towards physics, and in keeping their interest in physics alive throughout their school years. Moreover, the participants continued to employ the coping mechanisms detailed below throughout their lives, which afforded them advantages in their education and careers that helped to propel them to their current successes.

As a result of feeling isolated, the participants developed gendered identities which tended to be ambiguous, multiple, overlapping, and often contiguous rather than compartmented, and yet all were constructed within a heteronormative and traditional matrix. The outlier of this cohort claimed that she had close-knit friendships throughout primary school, but enjoyed working alone and did not take other people's comments seriously. Her narrative indicated that she developed the same coping mechanisms but did not have an exclusion experience similar to the other participants.

Chronologically, the participants' narrative accounts began by describing extended periods of social isolation, principally during their primary school years. According to the participants, they were outsiders, excluded from social groups both in and outside school:

> Diane: ...but no, some of it was a bit lonely probably, in retrospect, or maybe I was just always a loner, I don't know.
> Caroline: I never felt I belonged. At every stage in my life at school I was always much [different to] everyone else and also seen as teacher's pet. I was hated because I was teacher's pet and I was [different to] them so I was not interested in the same things.

The participants agreed that they were not members of the popular group at school. These perceptions of 'difference' to other children at school and of not fitting in seemed to result in the participants feeling lonely in their estimation. The participants' admissions of exclusion from peer groups in school often led to a discussion of their home environments. In addition to a lack of peer interaction in school, for some participants, feelings of isolation were exacerbated by contentious relationships with their siblings:

> Caroline: My sister was horrible to me.
> Faith: My younger sister suffered much more than I did from my brother. I was older and therefore had defence mechanisms, she didn't and my parents never noticed – they never knew. It's extraordinary actually. None of us ever told them, we just tried to cope – it was actually quite strange. I mean, they knew that he had a short temper, but they didn't know that he terrorized us.

For most of the participants who reported difficult family relationships in addition to social exclusion at school, their perceptions of isolation were the lenses through which they narrated their childhood. Three main types of defence mechanisms, or coping mechanisms, emerged from the participants' narratives; firstly, "developing a thick skin", secondly, becoming comfortable in working alone, and thirdly, acting less intelligent in order to fit in.

The first coping mechanism the participants cultivated in response to feeling excluded – developing a thick skin – could also be described as developing a high tolerance to the normalising discourse that surrounded them, as well as a greater reliance on polite social conventions to mask resentment or anger. The harassment that the participants experienced ranged from banter and teasing to a range of verbal, physical, and symbolic harassments like those tweeted as #everydaysexism ("Everyday Sexism Project," 2017). In response, instead of reacting angrily (or with embarrassment) to hateful banter or harassment, some of the participants seemed to be less emotionally sensitive to it, and consistently replied in a socially acceptable manner:

> Faith: A lot of my year group – friends if you'd like – would essentially tell me that if I wasn't prepared to announce my nationality, then I must be a terrorist. What else was there to discuss? And they would tell jokes that were only funny if you felt that [people of my nationality] were stupid and I had to learn to handle this – I didn't fly off. I learned to tell anti- English jokes and so on. But it reinforces outsider status…

Other participants learned to control their reactions to harassment so as not to evince any kind of physical or verbal response to antagonistic behaviours:

> Caroline: I don't remember crying myself to sleep or anything. I suppose you just take it as the norm. You just don't think. You just get on with it…

The participants perceived themselves as needing to cope with various forms of harassment by changing their behaviour and emotional responses; none indicated that they attempted to cope with peer or sibling bullying by preventing the episodes from occurring. As developing a thick skin is an emotional coping mechanism (Berger 2015), the participants appeared to be deeply emotionally affected by incidents of harassment, and yet learned to deflect their hurt. Instead of relying on the opinions of others, which were, according to the participants, often distressing or problematic, the participants began to rely on their own feelings and opinions:

> Diane: I've always had the mentality, the policy, that you think for yourself. You decide yourself whether you like something or you don't; you don't necessarily follow the crowds.

This reliance on the self was seemingly useful, in that it helped the participants to normalise their exclusion and gave them agency in their decisions (e.g. "you think for yourself", "you decide yourself"). Relying on their own opinions and being aware of the agency they had in their lives was mentioned as a turning point by most participants in the follow-up survey. The participants reported that, once they had reached this turning point, they decided that they were going to "do what they wanted [physics], no matter what popular stereotypes or other people said." Instead of responding to normalising judgement by conforming to the norm, the participants *accepted* their marginal status and began to identify as marginal, as the Other. However, their resistance to normalising judgement was more likely defensive rather than offensive: only after constructing a marginal identity post-isolation trauma were the participants able to identify as resistant to the norm.

At the same time, the participants suggested that they were only intimately exposed to one general performance of femininity on which they could learn to model their own behaviour: the femininity of the "mother." This feminine identity, extrapolated from participants' comments, was based on heteronormative and outdated versions of motherhood. The participants' resulting gendered identity was ambiguous and uncomfortable, constantly battling between remaining on the margin and tenuously creeping toward a more antiquated version of the norm. The second coping mechanism that emerged from the participants' experience of isolation was developing an ability, and for some a preference, for working alone. For some participants, working on their own began as a part of their isolating experience:

> Eve: I just got bunged at the back of the room with my own book because I was way ahead of everybody else, so I just, you know, did my own thing, which I was very happy to do.
> Interviewer: Were the other kids okay about that?
> Eve: Well, I think so. I mean, you had to learn not to boast and things. I think one learns to cope with being clever.

The participants reported that teachers and school administrators were aware of their exceptional intelligence, giving them advanced work or allowing them to take courses at nearby universities. In front of their peers, however, the participants performed an interconnected third coping mechanism—acting out a lesser intelligence to cope with their uniqueness. The participants were judicious in their performance of intelligence, in that they performed it in front of authority figures only, knowing that this would garner the most benefits. The terminology they used to describe being singled out to do advanced work (which reinforced their otherness) included positive terms, such as feeling "happy" or "wonderful." In fact, the participants reported that they continued to enjoy the work they did on their own from primary school all the way through to their positions as early career researchers. All the participants were involved with projects in laboratories, but also worked alone, and were very comfortable doing so. They described their work as "lovely," or "a wonderful time," but clearly indicated that most of their time was spent by themselves:

> Caroline: Well, a PhD is always on their own, technically. But, no, it was lovely. It was great and the technicians were very good too. They were very good looking after us. My relationship with my supervisor was... He was… a lovely chap but he was kind of not there a lot of the time. I remember, I mean - not there in the sense that if I wanted to talk to him, it could take a week to find him.
> Faith: Oh, no. I'm perfectly happy being alone. I don't mind company, but I also don't mind just me. So I didn't set out to avoid being lonesome.

Whilst studies have suggested that working alone can be damaging to mental health, the participants in this cohort seemed to thrive in that environment (Harnois and Gabriel 2000). This did not mean that the participants had no social interaction, however, but that their social interaction did not occur within educational, academic, or scientific milieu; the participants all sought social interaction within organised or institutionalised programmes, as will be discussed in the next section.

The defence mechanism of working alone allowed the participants to construct a resistant identity of social and intellectual independence—a masculine performance of autonomy and self-reliance in opposition to a heteronormative feminine performance of reliance on male suggestion or approval. This particular component of the participants' identities can be identified as "resistant" because they made an effort to justify independence as a positive trait. Moreover, they suggested that their isolation during their university years and beyond was voluntary and, at times, preferential to more integrated and collaborative working environments. The participants' choice to extend their isolation into their early careers implied a continued resistance to the norm, and to feminine performances of "social," "nurturing," and other qualities that are performed through interaction with others. The participants coped with their extraordinary intelligence by performing their intelligence judiciously (e.g. only for authority figures) which did not change others' opinions of them but kept them socially marginalized.

9.7 Alternative Ideologies

To navigate their feelings of isolation, participants turned to other arenas for social contact, support, and confirmation in their choice to become physicists. In their own estimation, the participants' lack in friendships at school was mitigated by relationships they formed in religious and music groups. In this way, both music and religion helped to construct the participants' earliest physicist identities which solidified their intention and validated their choice, whilst simultaneously making them believe that there was indeed a place where they belonged and were valued members of a group.

There was one outlier amongst the participants who did not report participating in either religion or music, but who was very involved with computer programming and computer games. As such, she was socially involved with other students who had similar interests, and in the 'B' group at school (not the popular group, but not the unpopular group). She reported having a rich social life, both in and out of school.

9.7.1 The Role of Religion

Three of the six participants were religiously observant, and referred to their beliefs, practices, or religious youth programs throughout the interview. Through their faith, religiously observant participants gained validation in what they believed was their purpose: physics. The denominations to which they belonged (two were Christian, one was Catholic) promoted a kind of faith-associated meritocratic ideology, which meant that they saw their skills and intellectual inclinations as holy gifts that should not be squandered:

> Faith: There was an attitude - which was very clearly stated – “you use the talent you’ve got” – God-given talents have to be used – so if you’re good at something you *should* be good at it. In that sense it was telling you to have aspirations.

The participants felt that their intellectual talents and passion for physics was encouraged by religious doctrine. Religion gave the participants a liberating message to follow their aspirations and pursue degrees in physics because they excelled at the subject. As their religious observance had been established early on in their lives (as a family activity and deeply embedded set of beliefs), the participants reported feeling more closely aligned with religious doctrine than normative (school peer/social) ideology. As the participants’ religious doctrine encouraged them to refine and pursue their God-given talents (interpreted by the participants as their talent for physics), they reported feeling validated in their choice of career path.

The participants did, however, express a contraindicative result by suggesting that their theology validated their masculine performances: for example, their God-given talent compelled them to participate in male-associated physics courses, and validated their marginalisation through the justification “God made me this way.” Whilst, in general, Christian theology is stalwart in its determination of gender roles, promoting a patriarchy based on reproductive possibilities, the participants chose to resist their doctrine-implied status and perform being a “master.” They did so by (re)positioning themselves within their theology to reflect an unmediated relationship to God, through which they could interpret God’s purpose for them without the intervention, negotiation, or interpretation of a (male) religious authority. The participants were again judicious, both in their understanding of theology and in their incorporation of particular aspects of the theology into their gendered identity. This alternative discursive formation allowed the participants to feel as though they were members of the norm even though their gender performance was, in some aspects, disruptive to the gender norms of their theology.

9.7.2 The Role of Music Groups

Three of the six participants were involved in musical groups that included orchestras, choirs, and English folk dancing. One respondent was both religiously observant and heavily involved in musical groups. Musical activities and clubs served as a safe space, especially for those participants who were religiously non-observant, where they felt accepted or judged based on their musical abilities rather than on their propensity to fit in:

> Caroline: And, I found friendships incredibly difficult. And that’s why music I think became very important at secondary school because in music [difference] wasn’t relevant. It was, you know, oh she is a [musician], good. I had an instant friendship group, as it were, and I think that was very important.

The participants' immediate incorporation into a group made them feel like they were a part of the norm, that they were not the deviants that normative ideology (such as that of their school peer groups) compelled them to believe they were. As in the religious groups, the primary reason for musical groups to meet was not necessarily social in the same way as friendship groups formed in school, but, instead, there was a clear agenda (playing a piece) and a goal (playing a piece [together] correctly/well). Moreover, the musical group was dependent on the participants' individual contributions and the participants were aware that this reliance gave them agency (to turn up, to play her piece well, to be in sync). Unlike their experiences in the classroom or in social environments, the participants were comfortable in musical groups and felt accepted:

> Caroline: I very rapidly got absorbed into all the music making which was in itself a problem because they were some very very good musicians and I was a beginner. But they accepted me…

It seemed that for the participants who were most actively involved in music groups, the comfort of being able to supply something that was in demand (i.e. an ability to play a particular instrument) was the impetus for joining and persisting in music groups. The participants felt as though their roles in music were clearly stated (e.g. "I played my part"), and that they were valued members of the group. Here, without the discomfort of social interaction (for the participants had already acknowledged their exclusion from social groups at school) the participants were able to become valued members of a group that supported who they were.

Some of the participants perceived their membership in music groups as being intertwined with social class. Two thirds of the participants who identified as musicians also identified as being working class, but as the participants became more involved with music, they perceived themselves as shifting towards the middle class. This, the participants reported, was helpful in encouraging them to persist in studying physics beyond secondary school. Their families had neither encouraged nor discouraged their intentions to study physics at university, as the participants' parents had not attended university. The participants reported imagining that they might have "gone down the same path", but felt that their involvement with music changed that:

> Anne: …You're almost automatically middle class because you have music every week, you sing a lot, you read at least one book on a regular basis, (both laughing) all that side of things is almost a cultural bridge to oh!, You read other books then, listen to other music. It does help you connect things into that whole circle.

This perception resonates with theories in music sociology, a field which built upon the work of Adorno and Bourdieu, amongst others, to investigate the ways in which musical taste and participation in musical groups and events evidence individual displays of social class (de Boise 2016). Shifting social class, or perceiving a shift in class, was beneficial in helping those participants to feel as though they fit in more during university (since they imagined higher education was a middle-class pursuit). This perception of being a member of the middle class through their engagement with music was likely crucial in supporting their intentions to pursue a degree and a career in science—neither of which have been highly correlated with working-class goals (Lucey et al. 2003). Becoming middle class through their own agency provided participants with a way in which they could fit in, even though they continued to stand out in other ways. That the participants continued to participate in music throughout secondary school, college, and university, whilst never considering it a possible career path, suggested that they wanted to reap the personal/social benefits musical groups provided. By engaging in the group and performing as a musician, the participants found a discursive moment in which they were considered a part of the norm. Having had this experience, the participants were (unconsciously) familiar with leveraging their own capital to progress in a discursive system. They used similar tactics to apply their science capital in a different discursive formation: physics. Without their participation in musical groups, the participants might not have chosen to participate in physics or persisted towards a degree.

9.8 The Working-Class Hero

Of the six participants, five self-identified as being working class, whilst the last described her grandparents as working class. None of the participants grew up with parents who had a university degree, and most described the world of academia as idealised in the notion of Oxbridge, which their parents and community considered alien to them:

> Eve: I mean, I think we were probably a family that was, it was a fairly working class background, you know … I remember someone got into Oxford who they happened to know and it was considered absolutely amazing, that Oxford was somewhere outside of the universe. Yeah, so, no. It never occurred to anyone that I was going to be anything but a, I guess a teacher, probably. A maths teacher.

Most of the participants equated science with Oxbridge, noting that both were normatively outside the realm of consideration for their families and schools. Further, all of the participants noted that they were amongst the first pupils at their schools to apply to Oxbridge or other Russell Group[1] universities:

> Beth: But, I guess I looked at Cambridge because in my sixth form college where they didn't really send very many people to [Oxbridge], they'd had more success with Cambridge, so they were like we don't really know any better, but more people seem to have got into Cambridge so maybe apply there.

Whilst one could be a success in academic physics if one earned an undergraduate degree from a non-Russell Group university, it does seem pertinent that all of the respondents attended Russell Group universities and most worked in them for their entire careers. Their affiliation with these universities seems to have added to their legitimacy and esteem, and yet, they began their association with

[1] The Russell Group Universities are a collective of 24 public research universities in the UK, arguably regarded as the best universities in the UK (https://russellgroup.ac.uk)

Oxbridge—coming from, what the participants described as, a minority group of working-class students. Moreover, they described the difficulties their parents had in finding funds to pay for their tuition, typically noting, "I mean we were never flush, but if it was for education we would do it" (Beth). The participants were never asked about class, but spontaneously referred to it when they constructed their background narratives, using specific details (such as scholarships, etc.) to highlight their disadvantages.

It could be argued that the participants' gendered identity was, on the surface, an inherent disadvantage to achieving success in academic physics, but the participants chose to emphasise their intersection between gender and class. The participants' perception of being doubly marginalised from the norm of physics (male and middle-class) created a narrative which made them seem especially exceptional and almost heroic: Coming from almost nothing, they had used their intellect and determination to succeed where few people like them ever had. The drama of this trope created a compelling narrative for the participants, a *Bildungsroman* of a sort, that allowed for more insight into their current gendered identities and their desire to break away from the working class. The participants clearly tried to perform a middle-class masculinity, performing intelligence, determination, and aspirations towards academic positions, for example, whilst suppressing yet maintaining their working-class femininity. In performing a middle-class masculinity, the participants shied away from laddish behaviours, and endeavoured to (re)create themselves based on middle-class masculine stereotypes about physicists. However, in addition to retaining ideas from childhood of women being mothers, the participants also struggled with working-class conceptions of women as mothers and housewives.

Through their non-verbal communications, the participants indicated with grimaces and defensive gestures that they were conflicted about marriage and motherhood and the expectations placed upon them by their backgrounds:

> Diane: … because of the way society was constructed at that time. I mean, I've already alluded to the fact that married women didn't work, but *mothers* didn't work because it was proven that if they did, their kid would be delinquent. So I was actually quite remiss putting him in care one day a month. (laughing)

As stated previously, half of the participants had divorced their first husbands, and their new partners were middle class. For participants like Diane, their childcare regime and position in the household changed dramatically with their change in partner, resulting in a more egalitarian state in which both partners took responsibility for the child. The participants indicated that they were happier with the newer childcare situation and attributed it to their partner's class.

By performing a middle-class masculinity, and suppressing their working-class femininity, the participants gained the agency to pursue a degree and career in academic physics. However, the gendered identity negotiations through which they laboured were difficult and disruptive. Through this study the participants recreated themselves as working-class heroes (even though they no longer had ties to the working class). Their identity work revealed both the struggles they endured and their passion to succeed.

9.9 Discussion and Conclusions

Determining why gender inequality endures in physics, and what can be done to ameliorate the situation, is a complex task. The research presented in this chapter approaches the chronic issue of gender equity in physics from a new perspective, by exploring the professional and personal pathways of some of the most successful female-identifying physicists in the UK and finding similarities between them. Firstly, the research revealed that participants developed a strong sense of self-reliance in their early experiences with isolation that allowed them to make their own educational choices, and to be less concerned with the opinion of others. Secondly, the participants' engagement in religious and musical support networks provided them with alternative ideologies to peer-enforced social ideologies. These networks reinforced the participants' desires to participate in physics and gave them the opportunity to fit into these contexts. Lastly, the participants (re)created themselves as working-class heroes, a process that acknowledged the intense identity work they struggled through in order to become who they are today.

All of their experiences and coping mechanisms provided insight into how they have become successful, and what might be done to widen participation for women in physics in the future. Similarities in their experiences provided insight into how women's participation and persistence in physics might be increased: through programs focusing on improving self-confidence and self-efficacy; by providing support networks; and in increasing and mandating workshops on unconscious bias and discrimination for all students, whilst maintaining positive actions for those identifying as female. In addition, the current system for promoting and awarding honours and distinctions appears to be meritocratic, which suited the participants. However, it might be more equitable in the future to standardise rewards, so that more women will apply and be awarded honours. For example, applying for the position of Professor might be more equitable if all faculty were required to do so after: (a) working in the field for a certain number of years; and (b) publishing a certain number of articles and book chapters. Whilst the participants in this study were able to achieve a certain normality within their departments, and were, therefore, recognised as physicists by their colleagues (and then nominated for promotions) it might be helpful for others to have a definitive checklist that would showcase their merit.

The participants in this study revealed that social mobility is possible through participation and persistence in academic physics. It may be useful to share these women's stories with working-class students to promote them as role models. Being aware that so many of the most successful women in UK academic physics are from working-class backgrounds may inspire working-class students to participate in physics.

Even though the above analysis discussed how performing masculinity helped the participants to persist in physics, it would be erroneous to presume that these findings indicate a need for those physicists identifying as female to shift their identify or perform exclusively as masculine. It is more the case that gender hybridity

(Danielsson 2014; Lucey et al. 2003), heteroglossia (Francis 2012), or identifying along the gender spectrum, should be recognised as the norm in physics. Thus far, achieving success has meant that those identifying or identified as women were required to negotiate their identities in order to fit in and garner the esteem they needed to receive grants and awards. However, broader non-dichotomous identity expectations in academic physics might be more effective in helping future generations to persist and find their own successes.

References

Archer, L., DeWitt, J., Osborne, J., Dillon, J., Willis, B., & Wong, B. (2012). Balancing acts: Elementary school girls' negotiations of femininity, achievement, and science. *Science Education, 96*(6), 967–989. https://doi.org/10.1002/sce.21031.

Archer, L., Osborne, J., DeWitt, J., Dillon, J., Wong, B., & Willis, B. (2013). *ASPIRES Young people's science and career aspirations, age 10–14*. Retrieved from http://www.kcl.ac.uk/sspp/departments/education/research/aspires/ASPIRES-final-report- December-2013.pdf.

Archer, L., Moote, J., Francis, B., DeWitt, J., & Yeomans, L. (2016). The "exceptional" physics girl: A sociological analysis of multimethod data from young women aged 10–16 to explore gendered patterns of post-16 participation. *American Educational Research Journal*. https://doi.org/10.3102/0002831216678379.

Baram-Tsabari, A., & Yarden, A. (2008). Girls' biology, boys' physics: Evidence from free-choice science learning settings. *Research in Science & Technological Education, 26*(1), 75–92. Retrieved from http://www.tandfonline.com/doi/pdf/10.1080/02635140701847538.

Berger, R. (2015). Challenges and coping strategies in leavening an ultra-orthodox community. *Qualitative Social Work, 14*(5), 670–686. https://doi.org/10.1177/1473325014565147.

Brickhouse, N. W., Lowery, P., & Schultz, K. (2000). What kind of a girl does science? The construction of school science identities. *Journal of Research in Science Teaching, 37*(5), 441–458. https://doi.org/10.1002/(SICI)1098-2736(200005)37:5<441::AID-TEA4>3.0.CO;2-3.

Butler, J. (1990). *Gender trouble: Feminism and the subversion of identity*. New York: Routledge.

Butler, J. (1993). Imitation and gender insubordination. In I. H. Abelove, M. A. Barale, & D. M. Halperin (Eds.), *The lesbian and gay studies reader* (pp. 307–320). New York: Routledge.

Carli, L. L., Alawa, L., Lee, Y., Zhao, B., & Kim, E. (2016). Stereotypes about gender and science: Women = scientists. *Psychology of Women Quarterly, 40*(2). https://doi.org/10.1177/0361684315622645.

Carlone, H. B., Webb, A. W., Archer, L., & Taylor, M. (2015b). What kind of boy does science? A critical perspective on the science trajectories of four scientifically talented boys. *Science Education, 99*(3), 438–464. https://doi.org/10.1002/sce.21155.

Creswell, J. W. (2003). *Research design: Qualitative, quantitative, and mixed method approaches*. Thousand Oaks: Sage Publications.

Creswell, J. W., & Creswell, J. W. (2013). *Qualitative inquiry & research design: Choosing among five approaches*. Thousand Oaks: Sage Publications.

Danielsson, A. T. (2012). Exploring woman university physics students 'doing gender' and 'doing physics'. *Gender and Education, 24*(1), 25–39. Retrieved from http://ezproxy-prd.bodleian.ox.ac.uk:2108/doi/abs/10.1080/09540253.2011.565040#.VMdzvVqZbww.

Danielsson, A. T. (2014). In the physics class: University physics students' enactment of class and gender in the context of laboratory work. *Cultural Studies of Science Education, 9*(2), 477–494. https://doi.org/10.1007/s11422-012-9421-3.

de Boise, S. (2016). Post-Bourdieusian moments and methods in music sociology: Toward a critical, practice-based approach. *Cultural Sociology, 10*(2), 178–194. https://doi.org/10.1177/1749975516628309.

DeWitt, J., Archer, L., & Osborne, J. (2014). Science-related aspirations across the primary-secondary divide: Evidence from two surveys in England. *International Journal of Science Education, 36*(10). https://doi.org/10.1080/09500693.2013.871659.

Dey, I. (1993). *Qualitative data analysis: A user-friendly guide for social scientists*. New York: Routledge.

Department for Education and Skills. (2007). *Gender and education: The evidence on pupils in England*. Retrieved from https://www.education.gov.uk/publications/eOrderingDownload/00389-2007BKT-EN.pdf.

Ecklund, E. H., Lincoln, A. E., & Tansey, C. (2012). Gender segregation in elite academic science. *Gender & Society, 26*(5), 693–717. https://doi.org/10.1177/0891243212451904.

Everyday Sexism Project. (2017). Retrieved September 18, 2017, from https://everydaysexism.com/

Francis, B. (2012). Gender monoglossia, gender heteroglossia: The potential of Bakhtin's work for re-conceptualising gender. *Journal of Gender Studies, 21*(1), 1–15. https://doi.org/10.1080/09589236.2012.639174.

Fuller, C. (2018). The existential self: Challenging and renegotiating gender identity through higher education in England. *Gender and Education, 30*(1), 92–104. https://doi.org/10.1080/09540253.2016.1241380.

Gill, P., Stewart, K., Treasure, E., & Chadwick, B. (2008). Methods of data collection in qualitative research: Interviews and focus groups. *BDJ, 204*(6), 291–295. https://doi.org/10.1038/bdj.2008.192.

Gonsalves, A. J. (2014). "Physics and the girly girl—There is a contradiction somewhere": Doctoral students' positioning around discourses of gender and competence in physics. *Cultural Studies of Science Education, 9*(2), 503–521. https://doi.org/10.1007/s11422-012-9447-6.

Gonsalves, A., Rahm, J., & Carvalho, A. (2013). "We could think of things that could be science": Girls' re-figuring of science in an out-of-school-time club. *Journal of Research in Science Teaching, 50*(9), 1068–1097. https://doi.org/10.1002/tea.21105.

Götschel, H. (2013). No space for girliness in physics: Understanding and overcoming the masculinity of physics. *Cultural Studies of Science Education, 9*(2), 531–537. https://doi.org/10.1007/s11422-012-9479-y.

Harding, S. (1993). Rethinking standpoint epistemology: What is strong objectivity? In L. Alcoff & E. Potter (Eds.), *Feminist epistemologies*. New York/London: Routledge.

Harnois, G., & Gabriel, P. (2000). *Mental health and work: Impact, issues and good practice*. Geneva. Retrieved from http://www.who.int/mental_health

Hazari, Z., Tai, R. H., & Sadler, P. M. (2007). Introductory university physics performance : The influence of high school physics preparation. *Gender and Education, 91*(6), 847–876. https://doi.org/10.1002/sce.

Higher Education Statistics Agency. (2016). *Data collection|HESA*. Retrieved from https://www.hesa.ac.uk/collection

Hughes, G. (2001). Exploring the availability of student scientist identities within curriculum discourse: An anti-essentialist approach to gender-inclusive science. *Gender and Education, 13*(3), 275–290. https://doi.org/10.1080/09540250120063562.

Johnson, A., Brown, J., Carlone, H., & Cuevas, A. K. (2011). Authoring identity amidst the treacherous terrain of science: A multiracial feminist examination of the journeys of three women of color in science. *Journal of Research in Science Teaching, 48*(4), 339–366. https://doi.org/10.1002/tea.20411.

Kantaria, P. (2012). *It's different for girls: The influence of schools A report looking at data from the National Pupil Database comparing boys' and girls' choices of science A-levels and the influence of their GCSE schools on these choices.* Retrieved from http://www.iop.org/education/teacher/support/girls_physics/file_58196.pdf

Lancaster, K. (2016). Confidentiality, anonymity and power relations in elite interviewing: Conducting qualitative policy research in a politicised domain. *International Journal of Social Research Methodology*, 1–11. Retrieved from https://doi.org/10.1080/13645579.2015.1123555

Lock, R. M., Hazari, Z., & Potvin, G. (2012). Physics career intentions: The effect of physics identity, math identity, and gender. In *Physics education research conference* (pp. 262–265). American Institute of Physics. https://doi.org/10.1063/1.4789702.

Lucey, H., Melody, J., & Walkerdine, V. (2003). Uneasy hybrids: Psychosocial aspects of becoming educationally successful for working-class young women. *Gender and Education, 15*(3), 285–299. https://doi.org/10.1080/09540250303865.

Miller-Friedmann, J., Childs, A., & Hillier, J. (2018). Approaching gender equity in academic chemistry: Lessons learned from successful female chemists in the UK. *Chemistry Education Research and Practice, 19*(1), 24–41. https://doi.org/10.1039/C6RP00252H.

Nosek, B. A., Smyth, F. L., Sriram, N., Lindner, N. M., Devos, T., Ayala, A., & Greenwald, A. G. (2009). National differences in gender-science stereotypes predict national sex differences in science and math achievement. *Proceedings of the National Academy of Sciences of the United States of America, 106*(26), 10593–10597. https://doi.org/10.1073/pnas.0809921106.

Phipps, A., & Young, I. (2015). Neoliberalisation and "lad cultures" in higher education. *Sociology, 49*(2), 305–322. https://doi.org/10.1177/0038038514542120.

Sadler, P. M., Almarode, J. T., Miller-Friedmann, J. L., Tai, R. H., Sonnert, G., Hazari, Z., & Dabney, K. P. (2012). Out-of-school time science activities and their association with career interest in STEM. *International Journal of Science Education, Part B.* https://doi.org/10.1080/21548455.2011.629455.

Sikora, J., & Pokropek, A. (2012). Gender segregation of adolescent science career plans in 50 countries. *Science Education, 96*(2), 234–264. https://doi.org/10.1002/sce.20479.

Sinnes, A. (2006). Three approaches to gender equity in science education. *NorDiNa, 2*(1), 72–83.

Stentiford, L. J. (2018). "You can tell which ones are the laddy lads": Young women's accounts of the engineering classroom at a high-performing English university. *Journal of Gender Studies.* https://doi.org/10.1080/09589236.2018.1423957.

Stewart, M. (1998). Gender issues in physics education. *Educational Research, 40*(3), 283–293. https://doi.org/10.1080/0013188980400302.

Venville, G., Rennie, L., Hanbury, C., & Longnecker, N. (2013). Scientists reflect on why they chose to study science. *Research in Science Education, 43*(6), 2207–2233. https://doi.org/10.1007/s11165-013-9352-3.

Chapter 10
Conclusions Part I: Responding to Frameworks and Methodologies that Attend to Gender in Physics Education: Practical Implications for Higher Education

Dimitri R. Dounas-Frazer

10.1 Introduction and Positionality

In their chapter, Diane Crenshaw Jammula and Felicia Moore Mensah (Chap. 5) demonstrate that physics students' subjectivities are dynamic and gendered, but not essential characteristics of their sex. Further, they argue that "physics teacher educators are tasked to broaden the ways that physics teachers think about physics and their students' multiple subjectivities" (p. 95). In her chapter, Angela Johnson (Chap. 4) describes a physics department in which the women students of color feel supported. In that department, "male physics faculty members take gender issues seriously, rather than leaving equity issues to their female colleagues". Accordingly, as a physics teacher educator and a male physics faculty member, I open my discussion by describing some of my own subjectivities, professional practices, and conceptions of physics. In doing so, I aim to provide context for, and thus facilitate criticisms of, my interpretations of the ideas in this book.

I am a white cisgender man, and my gender expression is typically interpreted as masculine. I am a former experimental atomic physicist and a current education researcher who studies teaching and learning in physics laboratory courses. I have been educated and trained in physics departments that are predominantly white and male, and I currently work in such a department. Similarly, I was raised in a predominantly white middle-class community, and I currently live in one. Thus, middle-class white masculinity has been a major socializing force in my professional and personal lives. This type of masculinity aligns with dominant conceptions of physics (Jammula and Mensah, Chap. 5), and it can cultivate a sense of entitlement or righteousness that facilitates injustice in the academy (Shahvisi 2015). To resist

D. R. Dounas-Frazer (✉)
Western Washington University, Bellingham, WA, USA
e-mail: dounasd@wwu.edu

A. J. Gonsalves, A. T. Danielsson (eds.), *Physics Education and Gender*, Cultural Studies of Science Education 19,
https://doi.org/10.1007/978-3-030-41933-2_10

my socialization into white masculinity, I recently entered into an accountability partnership with Regan Byrd and Simone Hyater-Adams, two Black women who have expertise in issues of race and gender. Our partnership draws on models for self-accountability that have been developed in antiviolence movements (Fujikawa et al. 2018). It provides structure, boundaries, and compensation for Byrd and Hyater-Adams to support me in processing feedback about my own racism and sexism in professional contexts and to set attainable and appropriate antiracist and antisexist professional goals for myself (Dounas-Frazer et al. 2018).

I am openly gay and queer. Queerness informs my knowledge of the world. For example, I am skeptical of dichotomies, including technical-social dualism. "Technical-social dualism" refers to the pervasive and oppressive belief that technical things are different from, and better than, social ones (Cech and Waidzunas 2011; Faulkner 2000). Taken to an extreme, this dualism can manifest as messaging that physics is more important than everything else. Indeed, some of the physics lecture jokes analyzed in this book convey precisely that message (Johansson and Berge, Chap. 6). Consistent with my aversion to binary thinking, my queerness implores me to view physics as both a powerful cis/heteronormative socializing force and a powerful metaphor for queerness and genderqueerness. As a physics student, some of my peers invoked physics models and apparatus (technical) to position my sexuality (social) as unnatural: opposite electric charges attract, identical charges repel, and prongs plug into sockets. As a more mature physicist, I now see myself and other gender/queer people reflected in the laws of the universe: quantum entanglement as metaphor for nonbinary and dynamic genders and quark confinement as metaphor for gender/queer solidarity. I am not alone in this queered conception of physics. For instance, Amrou Al-Kadhi, a queer nonbinary artist, has previously described how quantum physics helps them understand their own queer identity (Al-Kadhi 2018). The collective weight of these gendered subjectivities, experiences, and notions of physics informs which ideas and findings in this collection resonate with me most strongly and how those ideas could impact physics teaching practice in higher education.

10.2 Resonant Ideas and Findings

In their ethnographic description of students' identities and corresponding connection or disconnection to physics, Jammula and Mensah (Chap. 5) show that students' subjectivities are not essential: Naira, a Pakistani woman, sometimes enacted masculinity to defend her ideas; Sameer, a man from the Middle East, valued relationships and emotion, which are typically associated with femininity. That is, behaviors that are typically associated with whiteness or masculinity are not inextricably coupled to one's race or gender. As a white cisgender man who is attempting to resist the white masculine socializing forces of physics culture (e.g., elitism and hyperindividualism), this finding resonates with me because it gives me hope. It reinforces that other cisgender men and I are not biologically prohibited from enacting and modeling for our students behaviors that are typically coded as feminine (e.g., diligence and empathy). However, Jammula and Mensah also show that physics is aligned with middle-class white masculinity, and that gendered performances confer status in the classroom. Their work contains a message of caution: physicists of all genders who embrace (or aspire to embrace) feminine behaviors and values may experience a harsh disconnect with the dominant physics culture.

Due to my various professional identities, I am particularly interested in the performance and construction of gender in physics laboratory courses. For some students, laboratory courses can be formative experiences that position them as more central in a community of physics practice (Irving and Sayre 2014). Work in this book by Marianne Løken and Margareta Serder (Chap. 7) and Adrienne Traxler and Jennifer Blue (Chap. 8) suggests that laboratory courses may also be doing another kind of positioning. Løken and Serder (Chap. 7) employ a sociomaterial approach to illustrate that people's educational interests, aspirations, and choices are shaped in part by the "things that surround us, the experiences we have with them, and our bodily situation in the world". They describe how Mia and Violet, two women studying physics at a Norwegian institution, came to be interested in physics through formative childhood intra-actions with material experience: construction games for Mia and a rocket launch for Violet. This sociomaterial analysis is immediately applicable to laboratory courses, environments whose importance stems from the opportunity they afford students to use sophisticated physics apparatus (Caballero et al. 2018). Moreover, providing students with opportunities to design, build, and troubleshoot their own experiments can foster their sense of project ownership and their interest in corresponding physics topics (Dounas-Frazer et al. 2017b). Recent work by Allison Gonsalves, Anna Danielsson, and Helena Pettersson (2016) shows that "performances of masculinity in physics are constructed through tinkering with instruments designed for larger (male) bodies." (p. 020120–13). Therefore, analyses that consider the agency of equipment and software could provide crucial insight into students' negotiations of both their gender and their connection to physics through intra-actions with physics apparatus. That is, sociomaterial approaches like the one presented by Løken and Serder could help physics educators better understand our students' gendered material experiences in laboratory courses.

The social model of disability described by Traxler and Blue (Chap. 8) further complexifies the performance and construction of identity in laboratory courses. In their recent call to invest in the improvement of physics laboratory courses, Caballero et al. (2018) note that "labs may give rise to a unique combination of stereotypes, discriminatory behaviors, and mobility or sensory barriers that unfairly prevent full participation for some learners." Reinterpreting this call through the social model of disability, "barriers that unfairly prevent full participation" could be viewed as disabling structural features of the classroom. How might the type and use of equipment and software cause some learners to be disabled, regardless of whether they self-identify as having a disability? That is, how might the material and social conditions of laboratory courses contribute to a harsh disconnect with physics for students with certain types of body?

10.3 Practical Implications for Higher Education

In Chap. 2, Louise Archer, Emily MacLeod and Julie Moote suggest that redressing inequality in physics requires a disciplinary shift:

> [W]e suggest that the challenge (and potential) will lie in getting the field of physics (and the myriad of powerful actors within this field) to understand the ways in which social reproduction functions in this space – and to then accept a reduction in their previously-enjoyed privilege in order to genuinely redress the effects of inequality and to open up the field to a more diverse demographic of participants.

This broad call can feel overwhelming for individuals who want to take action in support of gender-, race-, class-, and ability-based equity and to eliminate the disconnect between physics and femininity. However, there are several concrete actions that men can take to disrupt inequities in physics. Focusing on higher education, I will draw on my own experiences as a white queer cisgender man who is involved in physics at four grain sizes: (i) professional society, (ii) university department, (iii) post-secondary classroom, and (iv) individual person.

Archer, MacLeod, and Moote call for increased understanding of mechanisms for social reproduction within the field. In higher education, one way that physics educators advance our collective understanding is through conferences. Conference sessions, panel discussions, and plenary presentations represent existing mechanisms that can continue to be leveraged to infuse new ideas and language about gender into the physics education community. Moreover, it is possible to leverage these mechanisms in ways that challenge the reproduction of middle-class white masculinity in physics. Invite experts from beyond the physics education community, and ensure that groups of experts are diverse with respect to gender, race, class, ability, and other dimensions of identity. Secure funds to defray speakers' travel costs or pay them honorariums. Advertise the event to ensure high attendance, and organize appropriate networking events for speakers and relevant community members. My experience is that women, especially women of color, are overrepresented among organizers of sessions, panels, and plenaries focused on doing gender in physics education. Thus, there is a need for men to take on the labor of organizing such events.

At the departmental level, Johnson described a department in which faculty use research-based teaching strategies, foster student collaboration, and view success in physics as the result of hard work. Importantly, male faculty members in that department "take gender issues seriously." If one is not already embedded in such a department, what can be done? Cultural change in physics and other science or engineering departments is an emerging area of focus in physics education. Corbo et al. (2016, 2018) have been employing a Departmental Action Team (DAT) model that involves teams of students, staff, researchers, professors, and trained facilitators working together toward enacting a shared vision for their department. Others have also used the strategy of assembling a team of students, educators, and facilitators external to the department. For example, in order to accommodate the specific needs of a blind physics major, one physics department assembled a team consisting of the student, a blind physics bachelor's degree recipient from a different institution, a consultant on accessible science education who is also blind, and multiple sighted faculty, staff, and students (Holt et al. 2019). Further, there is also a role for departmental investment in student groups dedicated to collective self-education about issues of equity in physics (Dounas-Frazer et al. 2017a). Physics faculty members, including and perhaps especially men, could familiarize themselves with this literature and advocate within their institution for resources to support cultural change through hiring of external facilitators or consultants; forming heterogeneous teams of faculty, students, and staff; and investing in student-led diversity organizations.

Whereas a single actor cannot easily change the department, the classroom is a space that can benefit from both collective and individual action. Classrooms are impacted by departmental commitments (or lack thereof) to using research-based teaching strategies and instructors' implementation of those strategies. Similarly, the notion that physics is elite can be reinforced or challenged in the classroom through structural gatekeeping practices (cf. Archer, MacLeod, and Moote, Chap. 2) and instructors' use of humor (Johansson and Berge, Chap. 6). After reading the other chapters in this book, I adjusted my large-lecture teaching practices to incorporate more explicit framing about the conditions for success in physics while also pushing back on technical-social dualism. I told my students that physics requires social, emotional, and communicative skills in addition to mathematical ones; that it is a collaborative human endeavor rooted in sociopolitical contexts; and that it is learned through practice and dialogue. Although these messages were constrained to the first day of class, there was a strong response. Two students told me they appreciated the messaging. Three others explicitly questioned my competence as an educator. I had never before received such criticism from even a single student, but my colleagues who are women found the criticism familiar. I wonder whether my sexuality and my teaching practices—in this case, the framing of just one lecture—worked together to construct me as sufficiently feminine that some students perceived me as misaligned with physics culture and therefore as having low status in the classroom (cf. Jammula and Mensah, Chap. 5). I further wonder about which social supports I can rely on as I continue to experiment with teaching framings or strategies that misalign with the dominant physics culture.

At the individual level, my accountability partners are an invaluable source of social support when it comes to challenging myself to take action in support of gender- and race-based equity in physics (Dounas-Frazer et al. 2018). This partnership is both proactive and reactive: my accountability partners help me set and reflect upon inclusive and attainable teaching goals and they help me anticipate and process my gendered and racialized experiences in physics and academia more generally. Archer, MacLeod, and Moote have called not just for changing physics itself, but also for a shift in how "powerful actors within this field" understand gendered socializing forces. Although the partnership I co-developed with Byrd and Hyater-Adams is new and experimental, I believe that this or similar mechanisms have the potential to facilitate a positive disciplinary shift through individual self-accountability.

References

Al-Kadhi, A. (2018). What quantum physics taught me about queer identity. *BBC Ideas*. (British Broadcasting Corporation, September 21, 2018). https://www.bbc.com/ideas/videos/what-quantum-physics-taught-me-about-queer-identit/p06ltbj1

Caballero, M. D., Dounas-Frazer, D. R., Lewandowski, H. J., & Stetzer, M. R. (2018) Labs are necessary, and we need to invest in them. *APS News: The Back Page*, 27(5). https://www.aps.org/publications/apsnews/201805/backpage.cfm

Cech, E. A., & Waidzunas, T. J. (2011). Navigating the heteronormativity of engineering: The experiences of lesbian, gay, and bisexual students. *Engineering Studies, 3*(1), 1–24. https://doi.org/10.1080/19378629.2010.545065.

Corbo, J. C., Reinholz, D. L., Dancy, M. H., Deetz, S., & Finkelstein, N. (2016). Framework for transforming departmental culture to support educational innovation. *Physical Review Physics Education Research, 12*, 010113. https://doi.org/10.1103/PhysRevPhysEducRes.12.010113.

Corbo, J. C., Quan, G. M., Falkenberg, K., Geanious, C., Ngai, C., Pilgrim, M. E., Reinholz, D. L., & Wise, S. (2018). *Externalizing the Core principles of the departmental action team (DAT) model. Proceedings of the 2018 physics education research conference*. Washington, DC: AAPT. https://doi.org/10.1119/perc.2018.pr.Corbo.

Dounas-Frazer, D. R., Hyater-Adams, S. A., & Reinholz, D. L. (2017a). Learning to do diversity work: A model for continued education of program organizers. *The Physics Teacher, 55*(6), 342–346. https://doi.org/10.1119/1.4999728.

Dounas-Frazer, D. R., Stanley, J., & Lewandowski, H. J. (2017b). Student ownership of projects in an upper-division optics laboratory course: A multiple case study of successful experiences. *Physical Review Physics Education Research, 13*(2), 020136. https://doi.org/10.1103/PhysRevPhysEducRes.13.020136.

Dounas-Frazer, D. R., Byrd, R., & Hyater-Adams, S. (2018). A model for self-accountability in academia, physics education research consortium of graduate students. *Newsletter, 14*, 4.

Faulkner, W. (2000). Dualisms, hierarchies and gender in engineering. *Social Studies of Science, 30*(5), 759–792. https://doi.org/10.1177/030631200030005005.

Fujikawa, K., Perez-Darby, S., & Kaba, M. (2018). *Building accountable communities*. Barnard Center for Research on Women. (Barnard College, October 26, 2018). http://bcrw.barnard.edu/event/building-accountable-communities

Holt, M., Gillen, D., Nandlall, S. D., Setter, K., Thorman, P., Kane, S. A., Miller, C. H., Cook, C., & Supalo, C. (2019). Making physics courses accessible for blind students: Strategies for course administration, class meetings, and course materials. *The Physics Teacher, 57*(2), 94–98.

Irving, P. W., & Sayre, E. C. (2014). Conditions for building a community of practice in an advanced physics laboratory. *Physical Review Special Topics – Physics Education Research, 10*, 010109.

Shahvisi, A. (2015). Epistemic injustice in the academy: an analysis of the Saida Grundy witch-hunt. *Academe Blog*. (Academe Magazine, May 20, 2015). https://academeblog.org/2015/05/20/epistemic-injustice-in-the-academy-an-analysis-of-the-saida-grundy-witch-hunt

Chapter 11
Conclusions Part II: Implications of Identity Research for Upper Secondary Educators

Christopher Gosling

11.1 Background

There is enormous potential for using identity as a tool to examine how students learn physics and position themselves (and each other) relative to the field. Before launching into a discussion of the implications this collection holds for upper secondary educators, I will first describe my own background and context so that readers can better understand my perspectives.

To begin with, I identify as a White, cisgender, middle-class man, which means that when students look at me they see the dominant stereotype of what physicists are presumed to look like. I completed a bachelor's degree in engineering and worked as a structural engineer before switching careers to become a teacher. My professional interests as a teacher and graduate student lie in issues of underrepresentation in physics, namely around issues of gender. I have taught at both the secondary and post-secondary levels in the past and currently teach in a rural public high school in the northern part of New York in the United States. The school where I teach is diverse from a socioeconomic perspective but not racially: the student body of ≈ 350 is 98 + % White.

During my tenure at this school I have taught a variety of courses in physics, science, mathematics, and engineering. The mainstay of my teaching load has been an introductory algebra-based physics class, which approximately 50% of students elect to take. The lab component of this course is largely comprised of independent authentic experiments which students design and run parallel to class. Also, during this course my students and I explore issues of underrepresentation and have difficult conversations about race, gender, and privilege.

C. Gosling (✉)
Saranac Lake High School, Saranac Lake, NY, USA
e-mail: cwgosling8@gmail.com

A. J. Gonsalves, A. T. Danielsson (eds.), *Physics Education and Gender*, Cultural Studies of Science Education 19,
https://doi.org/10.1007/978-3-030-41933-2_11

11.2 Lessons

In this chapter I will review this collection with a specific eye for how the ideas these scholars present can be applied to settings similar to my own. While I am a secondary educator, I will also suggest applications for the elementary and middle-school contexts. Physics is often talked about in ways that preserve its elite status and reputation as being only for certain types of people, which shows why we must change how science is presented to students throughout the K-12 spectrum.

I will begin with a simultaneous consideration of the two chapters written by Louise Archer, Emily MacLeod, and Julie Moote. Both pieces used a Bourdieusian conceptual lens to gain insights into how students experience physics in gendered ways. In Chap. 3 the authors described physics habitus as: "a configuration of socialised dispositions that are structured and cultivated by the field of (school) physics which... structure young people's views and experiences of the subject and shape the extent to which they feel it is 'for me' or not" (Chap. 3, p. 34). This perspective takes a sociocultural approach to learning, which entails not only an exploration of how students learn content, but also how they navigate the structures of the local classroom culture (Lemke 2001). These chapters demonstrate a mismatch between how students see themselves and the way that the field of school physics asks them to behave. As the authors suggest, this frequently has powerful ramifications for students' identity and career trajectories.

In Chap. 2, Archer, MacLeod, and Moote discuss why mismatches between how students saw themselves and how they were positioned by the field negatively impacted their trajectories. For example, Danielle's school did not allow her to study A-level physics and both Thalia and Victoria were forced out at the end of their first year of A-level physics. It is important to emphasize that these departures happened despite the girls being very interested in physics. Moreover, the gatekeeping strategies employed by the schools and teachers resulted in the students placing blame on themselves for their so-called "failure" rather than viewing the system as complicit.

We see similar patterns in students who finished A-level physics. Despite completing this rigorous course of study, Davina, Kate, and Mienie were positioned as outsiders by the field of physics. Specifically, since they did not align themselves with the notion of the "'effortlessly clever physicist'" (Chap. 2, p. 14) which is traditionally associated with masculinity, these three students "came to the view that degree level physics was not 'for me' [them]" (Chap. 2, p. 14). This sentiment was also echoed by Hannah, who successfully completed A level physics. Though she continued in physics, she also did not align herself in non-dominant ways. For example, the authors describe how Hannah viewed herself as being "'good at physics'… via feminised diligence" (Chap. 2, p. 18). This position stands in opposition to the traditionally masculine identification patterns of physicists as being effortlessly clever.

As I read this chapter, I found myself curious as to why Hannah alone was ultimately able to succeed. Archer, MacLeod, and Moote tackled this difference head-on, first exploring the possibility that Hannah's perseverance might be related to her substantial *physics*-related capital. The authors concluded that such capital alone could not explain her persistence, and turned to class differences to explain her distinct trajectory. They suggested that Hannah's upper middle-class background prepared her for entry into an elite field such as physics, meaning that her entry required her to navigate gendered norms but not classed norms. I do not mean to minimize the substantial work that is required to successfully do so, nor the significant compromises that Hannah made to her femininity in order to be recognized as an intelligible physicist. Rather, I seek to point out that Hannah's advantage was that this was the primary hurdle she faced: other girls described in this study also had to learn to navigate elitism *in addition to* navigating the gendered behaviors required of them by the field of physics.

The experiences of the girls in Chap. 2 stand in contrast to the subject of Chap. 3: a boy named Victor. Victor had much in common with these girls: an intense interest in physics and the competence necessary to succeed in the field. However, he had an additional asset in the form of a naturalised physics identity which was confirmed and supported by his family. Furthermore, Victor's ways of behaving were aligned with the field of physics. His "'geeky masculinity'" was consistent with the behaviors expected by physics. In addition, while Victor had to do substantial work to maintain his position as a "clever" student, this position allowed him to enter the field with a physics identity that was aligned with dominant norms. Victor walked through the door with a body that was readily accepted and an outlook that positioned him as an insider to the field. In the following section I will describe how educators can create spaces that are easier for non-dominant students to navigate.

11.3 Implications

Hazari et al. (2017) found that secondary physics teachers have an enormous effect on the formation of students' identities relative to physics. However, such impacts are not always positive. As Carlone and Johnson (2007) found, established members of the field recognize students' competence in various ways, which can either promote or discourage the formation of their science identities. The chapters in this book give insights into how secondary educators may support non-dominant students' identities in physics.

11.3.1 Cleverness

A pervasive theme in Chaps. 2 and 3, as well as in Traxler and Blue's Chap. 8, is that physics is regarded as a field suited for people who are clever or brilliant. This is not new: Traweek (1988) chronicled such attitudes in particle physics more than

20 years ago (p. 79), and this view extends to popular culture (e.g. "*The Big Bang Theory*", as discussed in Chap. 2). What Chaps. 2 and 3 offer is insight into how the stereotype of the clever physicist impacts non-dominant students: they are positioned as "other" and have immense difficulty forming a physics identity if they cannot reconcile how they see themselves with such expectations. Archer, MacLeod, and Moote demonstrated how this perception can be reinforced by teachers, who may attribute achievement in gendered ways. The solution to these tendencies, which are driven by our unconscious biases, is not to praise everyone in the same way. Rather, our goal should be to provide meaningful recognition (Hazari and Cass 2018) based on the nuances of the student and context.

One technique I have used to help students combat outsider feelings is to help them develop an awareness of how they perceive themselves in relation to others. I used this move earlier this year with a senior physics student who positioned herself as inferior to some of her classmates, who she referred to as "geniuses." I asked her why she thought that, and she replied that she had to work a lot outside of class and review her notes as she was studying to keep up. In response I pointed out that, based on class discussions: (a) she was connecting more deeply than her peers were and could solve problems in multiple ways, and (b), her test scores were near perfect. While these are not the only metrics of success, this conversation gave me the opportunity to recognize her as an intelligible physicist face to face.

Teachers also interact with students non-verbally via written feedback on assignments. I try to examine and limit my own biases when marking student work. The first step I take is to go through all the papers and flip to the second page so I can't immediately see the student's name. Then I shuffle the papers and begin marking. I score all the students' responses to a given page before moving to the next page. While I still recognize some students' handwriting, this process helps ensure that the feedback I'm giving is based on the work that the students did rather than their gender or my perception of how they *ought* to do. I still see the students' names on the first page (which I mark last), so this gives me the opportunity to add personal feedback and position them in constructive ways.

As teachers, we have the capacity (and duty) to counteract myths such as the "effortlessly clever physicist", and this work needs to start early on. This immediacy can be seen in the Chap. 3 example of Victor, who at the beginning of the study stated that anyone could succeed in physics, and gradually came to see it as a field which required a particular level of intelligence to enter. Clearly then, teachers need to work to actively dispel the myth that one can be successful in physics only if it comes easily, but how can this be achieved?

One technique I have used to this end is the sharing of stories of struggle with my students. In my context, students are generally expected (but not required) to take a standardized state exam at the end of the year. Students enter the class preoccupied with the exam- it is the most frequent response to the question "What about this class worries you?" which is included in a questionnaire I give my students for their first homework assignment. Students in lab were recently discussing the exam and the implications of failing it. I addressed the issue directly, relating the story of a former student who failed the exam and is currently working on a PhD in astrophysics. We looked up their picture and CV, which showed these students how a trajectory in physics is possible despite setbacks.

11.3.2 Gendered Norms

In Chap. 2, Archer, MacLeod, and Moote noted that Hannah was able to persevere through A-level physics because she "eschew[ed] popular femininity in order to be intelligible as a physicist" (p. 21). The need for non-dominant students to modify their behaviors to fit the expectations of the field shows that the problem lies with the field and not the students. Teachers need to work to actively dispel the myth that physics is a masculine field (Francis et al. 2016) and that physicists must behave and dress in particular ways in order to be taken seriously.

I learned to work against these expectations through the Underrepresentation Curriculum Project (Rifkin 2016). I begin with an activity where students describe what a physicist looks like and collectively searching for commonalities in students' descriptions. The stereotypes that appear can then be refuted by having students research working physicists who identify in non-dominant ways and sharing their findings with the class to help dispel preconceived notions about "who" can succeed in physics. I recognize that since I am a member of the dominant group in physics, I may more readily embrace such conversations than non-dominant physics teachers, who may themselves struggle to be recognized as intelligible physicists by their students.

This year my students' descriptions of physicists were particularly powerful. I am aware that students in the rural school where I teach generally have little exposure to physics before taking my class. However, this year students confessed that some of the descriptions generated during this activity, such as White men wearing glasses and dress shirts, were based on my appearance. This was a revelation for me: while I long since realized that I embody many of the characteristics of a stereotypical physicist, I didn't realize the extent to which some students view *me* as the paradigm for what a physicist looks like. This moment made me realize how important this activity was for my students to explore what other physicists look like. I built on this experience by inviting a woman who works as a postdoc in physics education to visit. I included her in class discussions and positioned her as an expert, asking her to weigh in when a lab group was unsure as to what was going on with the experiment they designed. The students responded positively to her explanation and later included it in their presentation, taking care to attribute to her using her appropriate title (Dr.).

11.3.3 Interest

A third theme that resonated with me from these chapters is that the underrepresentation of girls in physics has nothing to do with the girls themselves and everything to do with the field of physics. This is seen clearly in the experiences of the girls documented in Chap. 2, most of whom were either excluded from A level physics or positioned as outsiders by the dominant discourses of the field. The takeaway, which runs contrary to many well-intentioned initiatives, is that we do not need to work to get the girls interested in physics. Girls are likely *already* interested in physics! Rather, teachers need to work to keep the norms of physics from positioning non-dominant students as outsiders and to work against gatekeeping mechanisms which frequently prevent non-dominant students from pursuing their interests in the subject.

What does this look like on the ground? A starting point would be to address the prerequisites for physics. For example, in my school there is an unwritten understanding that students will not be ready to take physics until 11th grade because they won't have learned the necessary math until that time. The fallacy of this reasoning is that I reteach the mathematical skills necessary for physics each year. There are several reasons for this. Some students, despite being in 11th grade, are still in lower level math courses. Others have advanced in their coursework but are not comfortable deploying the skills needed to solve problems. Articulating this argument is one thing, but convincing the people who craft students' schedules is another task. I strongly encourage teachers, especially at the secondary level, to begin the conversations.

11.4 Conclusion

Archer, MacLeod, and Moote have shown that physics needs to be transformed to combat its elitist status (Chap. 2). I ask myself how I can help as a secondary physics teacher. In addition to the steps I outlined above, I suggest that teachers constantly examine their policies and actions with a critical eye for the messages that students are receiving. To quote Angela Johnson: we must ask "What do personal interactions, cultural features and structures in the setting convey about what kinds of people belong here?" (Chap. 4, p. 55). We must be aware of the culture that we are enculturating students into, and that culture begins with us. If the field of physics is to more readily recognize non-dominant students as potentially intelligible physicists, it falls to us to teach our all of our students' behaviors that support this movement.

References

Carlone, H., & Johnson, A. (2007). Understanding the science experiences of successful women of color: Science identity as an analytic lens. *Journal of Research in Science Teaching, 44*(8), 1187–1218. https://doi.org/10.1002/tea.20237.

Francis, B., Archer, L., Moote, J., DeWitt, J., MacLeod, E., & Yeomans, L. (2016). The construction of physics as a quintessentially masculine subject: Young people's perceptions of gender issues in access to physics. *Sex Roles, 76*, 1–19.

Hazari, Z., & Cass, C. (2018). Towards meaningful physics recognition: What does this recognition actually look like? *The Physics Teacher, 56*(7), 442–446. https://doi.org/10.1119/1.5055325.

Hazari, Z., Brewe, E., Goertzen, R. M., & Hodapp, T. (2017). The importance of high school physics teachers for female students' physics identity and persistence. *The Physics Teacher, 55*(2), 96–99. https://doi.org/10.1119/1.4974122.

Lemke, J. L. (2001). Articulating communities: Sociocultural perspectives on science education. *Journal of Research in Science Teaching, 38*(3), 296–316. https://doi.org/10.1002/1098-2736(200103)38:3<296::AID-TEA1007>3.0.CO;2-R.

Rifkin, M. (2016, January 29). Addressing underrepresentation: Physics teaching for all. *The Physics Teacher, 54*(2), 72–74. https://doi.org/10.1119/1.4940167.

Traweek, S. (1988). *Beamtimes and lifetimes: The world of high energy physicists*. Cambridge, MA: Harvard University Press.

The manufacturer's authorised representative in the EU is Springer Nature Customer Service Centre GmbH, Europaplatz 3, 69115 Heidelberg, Germany. If you have any concerns regarding our products, please contact ProductSafety@springernature.com

Printed and bound by CPI Group (UK) Ltd, Croydon, CR0 4YY
05/10/2026
02228121-0008